GUIDED MISSILE FRIGATE TROMP

REPLACING THE CRUISERS

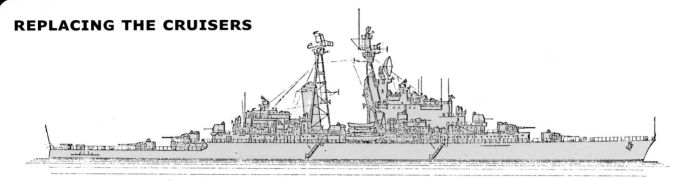

HNLMS *De Ruyter* and *De Zeven Provinciën* were the last cruisers of the Royal Netherlands Navy. In a period most ships were transferred from abroad (UK and USA), they were the largest post-war naval ships of Dutch manufacture. For years they were besides aircraft carrier *Karel Doorman* flagships. Construction of both ships started before World War II, but they did not enter service until 1953. After twenty years of service they were sold to Peru.

In May 1973 *De Ruyter* was renamed *Almirante Grau*. Modernizations 1985-88 and 1993-96. Decommissioned September 2017 (served 44 years with the Armada Peruana) to become a museum ship.

In August 1976 *De Zeven Provinciën* was renamed *Aguirre*. RIM-2 Terrier SAM removed, replaced by a hangar with large flight deck for three ASH-3D Sea King helicopters. Decommissioned 1999.

Cruisers De Ruyter and De Zeven Provinciën (in 1953)	
Displacement	11,850 tons (fl)
Dimensions oa.	*Length:* *De Ruyter* 187.32 m (614½ ft) had Atlantic (clipper) bow *De Zeven Provinciën* 185.70 m (609 ft) Beam: 17.25 m (56½ ft) Draught: 6.40 m (21 ft)
Machinery	4 Werkspoor-Yarrow three-drum boilers 2 De Schelde Parsons geared steam turbines 85,000 shp
Max. Speed	32 kts
Complement	926 -973
Armament	8 × 152 mm twin turrets, 1942 model Bofors 8 × 57 mm AA guns (4x2) Bofors 8 × 40 mm AA guns Bofors 2x DC racks 1 x 10.3 cm illumination rocket launcher

TROMP

INTRODUCTION

In 1964 new plans were developed concerning the structure of the fleet within the first six years of the seventies.
The intention was to decommission the carrier and replace the cruisers by two or four guided missile frigates. They would be equipped with an automated force AAW weapon-system TARTAR. Their coordination system consisting of the 3D radar and an automatic Combat Information Processing and Distribution System (DAISY)* with automated inter-ship data-links. In October 1970 an order was placed with KM De Schelde in Vlissingen (Flushing) for the delivery of two GM**-frigates.

*   The Tactical Data Handling System was called **DAISY,** the acronym stands
    for: **D**igitaal **A**utomatisch **I**nformatieverwerkend **SY**steem

** GM-frigates = Guided Missile frigate (FFG),
    in Dutch: GW-fregatten, GW = Geleide Wapen

Guns or missiles?

A fundamental change was the rise of self-propelled missiles, which alter the relationship between the power of the weapon and the demands it placed on the launching ship. Self-propulsion eliminates the need for elaborate launching equipment (i.e. heavy guns) and recoil effects. It is fair to put that the balance of costs shifted from a relatively inexpensive round fired by an expensive weapon equipped with an elaborate fire control system, to the opposite: an expensive single round requiring, often, rather inexpensive investment in acquiring launcher and fire control (upon the extent to which the missile is self-guiding).

Missiles provide small warships with the firepower of the capital ship of the past.

Impression drawn by C. Olling (1973). Note the different shape of the large radome.

Short response time became necessary. The new threat required changes in the build up of the fleet and its armaments.
A decision had to be taken to modernize or replace the large ships of the fleet, the latter being chosen for cost reasons. Technological developments also played a role. In the new design automation was saving space. The development of gas turbines for propulsion was one of these. It resulted in a personnel reduction. Gas turbines were immediately operational and increased readiness (not raising steam). The machinery was remote controlled.
The development of a 3D radar in combination with an automatic combat information system (DAISY) was another innovation that appealed to the Royal Netherlands Navy. With the 3D radar, it became possible to establish, besides bearing and distance, also the altitude of incoming objects in the air and report these contacts to fire control (WM-25).

June 1986. Early in the misty morning dressed for the Navy Days.

To answer the threat, Cold War in the sixties

By mid-sixties the Soviet threat was twofold. Soviet ballistic missiles and cruise missiles could be launched by submarines. While the first were targeting on land, the second could be used against ships. A Soviet submarine could store 4 to 8 missiles but needed to surface about 10 minutes for launching. The known missiles had a range of 350 and 650 miles and carried nuclear or conventional charges of 1000 to 2000 pounds. Most threatening to ships were cruise-missiles at an altitude of 1000 to 3000 feet. Only when the missile came within 10 miles range it could be detected by the ships radar. Leaving 60 seconds to react!
Requiring earlier detection and rapid reaction. New generation refined radars and electronics had to be developed.

Under the Forth Bridge across the Firth of Forth.

Predecessors of Tromp

1 1777 - 1796

A 54 gun ship of the line of the Admiralty of the Maze (Meuse) in Rotterdam. Its full name was *Maerten Harpertz Tromp,* joining the fleet in 1782.
In 1784 some German ships tried to enforce the river Scheldt. Being on Flushing roads the ship intercepted and challenged the intruders. Appointed to a task group and sent to Netherlands Indies in February 1796. These ships surrendered to the British on 17 August 1796. *Tromp* was confiscated and enrolled in the Royal Navy. Sold for scrap in 1813.

2 1804 - 1826

Gunboat / schooner number 22 carried 7 guns, built as *Admiraal Cornelis Tromp* in 1804. The ship took part in countering the Walcheren Campaign of the British in 1805. Sold for scrap in 1826.

Tromp 1877-1904. Dutch classification Screw Steamship 1st Class. In May 1893: Frigate.

3 1808 - 1823

Ship of the line of 64/68 guns *Maarten Harpertszoon Tromp*. Laid down in 1808 and commissioned on 5 May 1811. Departed Flushing on 16 March 1817. Sailing via Rio de Janeiro to Netherlands Indies, arriving 13 September at Anjer. Joining the expedition against the British lieutenant-governor Sir Thomas Stamford Raffles who still occupied the north of Sumatra. On 28 November 1818 the British handed over the area. In 1819 *Tromp* joined the expedition against Banka and in 1820 against Palembang. Later that year she was considered unfit for service and transferred to the Colonial Navy. Scrapped in 1823.

When the ship arrived, the situation was already under control. In 1885 she sailed to the Indies. Home bound she recorded the fastest speed ever made by a R.Neth. Navy ship. Between 1888 and 1890 journeys were made to Norway, South-Africa and West Indies. Again to the East in 1893.
Involved in various operations and returned in 1897. In 1899 she was appointed flagship of the Atjeh-division until this was dissolved in January 1902. Before returning the ship transferred 6 guns, ammunition and stores to establish a stronghold. She returned to Den Helder in 1902 and sold for scrap in 1904.

4 1850 -1867

A 74 gun ship of the line. Although the construction started in 1830, it took 20 years before launching. The ill fated project came to an end in 1867. The ship never sailed and was scrapped in 1872.

6 1904 - 1933

Coast Defence Ship *Marten Harpertszoon Tromp* was built at the Rijkswerf in Amsterdam. Commissioned on 5 April 1906. In June she sailed to Norway to witness the coronation of King Haakon VII. Departed in 1906 to Netherlands Indies where she was flagship of the Indies squadron. In April 1908 she sailed with the squadron to suppress armed resistance in Kloengkoeng (Bali).

5 1877 - 1904

Ship-rigged unprotected cruiser with sheathed iron hull built in Amsterdam. *Tromp* was commissioned on 30 May 1879. Leaving in October 1882 for Netherlands Indies, returning one year later. While sailing home she received orders to head for the Kongo river estuary where native Africans attacked a Dutch trading station.

Right: Marten Harpertszoon Tromp 1904-1933.

Sailed in 1909 in company with *Koningin Regentes* and *De Ruyter* to China and Japan before returning to the Netherlands. Late 1910 again employed in the Netherlands Indies. Enforcing neutrality until 1917.

Her last turn in the East was from 1919 to 1922. Once returned she became an instruction vessel and several training journeys were made to the Baltic, Canary- and Mediterranean waters. Decommissioned for the last time in 1927 and sold for scrap in 1933.

Tromp 1937-1969

7 1937 - 1969 (Warship No. 1)

Light cruiser, commissioned in 1938. Throughout 1940/41, she carried out patrol and escort duties with the Netherlands East Indies Squadron. Following the outbreak of WWII in the Pacific, she was assigned to the Combined Striking Force, ABDA Command. On 18 February 1942 *Tromp* was badly damaged off Bali (Badung Strait) and directed to Australia for repairs. Convoy duties. In February 1943, assigned to the US Seventh Fleet. In January 1944, *Tromp* joined the British Eastern Fleet based at Colombo (Ceylon), and stationed at naval base Trincomalee. Carried out raids on Sabang in April and Surabaya in May 1944. In the final months of the war, *Tromp* provided gun fire support preceding Allied landings at Balikpapan to recapture Borneo. Joined the British Pacific Fleet and in September 1945 deployed to Batavia where she landed marines who re-occupied the governor's residence. She remained in Sydney until February 1946 when she sailed for the Netherlands to repatriate ex-POW's. Arriving in May 1946, the ship underwent a significant refit which lasted until mid-1948. From 1949 onwards *Tromp* was a training ship. Since 1955 accommodation vessel, stricken from the list 1968. In 1969 sold and broken up.

8 1975 - 1999 (Warship No. 12)

Subject of this book.

Maerten - Marten - Maarten
His Christian name was "Maerten", sometimes written "Marten" but in the last century more often "Maarten"

9 2003 -

Frigate of *De Zeven Provinciën* class, commissioned 14 March 2003.

As easy to recognize the ships often used to promote the navy. Some stickers that appeared over the years.

HR.MS. TROMP - HR.MS. DE RUYTER

KONINKLIJKE MARINE

NATIONALE VLOOTDAGEN

KONINKLIJKE MARINE

Navy Days 1984 and 1989 (right)

NATIONALE VLOOTDAGEN

Den Helder
30 juni 1 en 2 juli
1989
NO 24

KONINKLIJKE MARINE

TROMP KLASSE
F 801
KONINKLIJKE MARINE

KONINKLIJKE MARINE

1988 / 1989

3D MTTR (multi target tracking) Long range air surveillance. Range 390 km (210 nm) Frequency Band: S. Weight aerials in radome: 3500 kg (7700 lbs). Basic architecture of the rotating antenna system consists of two parabolic search aerials (back to back) and two planar aerials (back to back) for the tracking function. 3D radar data was processed by computer and projected on screens.

Technical data	
Displacement:	3,724 tons standard / 4,377 tons fl
Length:	138.4 m (454 ft)
Beam:	14.8 m (48 ft 7 in)
Draught:	4.6 m (15 ft 2 in)
Machinery:	COGOG: 2x Rolls-Royce Olympus TM3B gas-turbines (44.000 shp (32.4 MW)) 2x Rolls-Royce Tyne RM1A gas-turbines (8.000 shp (5.9 MW))
Speed:	2 shafts, 28 kts maximum / 18 kts cruising
Range:	5,000 nm (9,300 km) at cruising speed
Complement:	306 (1975)
Armament:	1 × Mk 13 launcher for Standard SAM (40 missiles) 1 × Mk 29 launcher (octuple) Sea Sparrow SAM (16 missiles) 8 × Mk 141 launcher for RGM-84 Harpoon SSM 2 × 12 cm Bofors guns, twin turret 2 × Mk 32 triple anti-submarine torpedo tubes (324 mm) 1 × 30 mm Goalkeeper CIWS (1990 *De Ruyter* / 1995 *Tromp*) 2 x 20 mm Oerlikon gun
Helicopter:	1 × Westland WG-13 Lynx

By 1965 the Royal Netherlands Navy announced the intention to build new ships. The concept called 'Frigates 1965' following a design by Ir. J.E. Los of the department 'Materieel van het ministerie van Marine'. The project-coordinator was Captain G.W.A. Langenberg. Early 1970, the specifications were completed and offers were requested from the shipyards Rijn-Schelde and Verolme. This yielded an embarrassing situation because the R. Neth. Navy had already granted the order to N.V. Koninklijke Mij. De Schelde (KMS) in 1969. The Verolme yard was at that moment already extensively investigated and it was concluded that the construction with KMS would be associated with fewer risks. Understandable Verolme squabbled and questioned the decision, but in vain: both ships were eventually built at the KMS in the course of 1973-1974 launched and entered service two years later.

The new ships were specialized in air defence and easy to identfy by their large 3D radome. In addition, the 12cm Bofors turrets formed a striking feature.

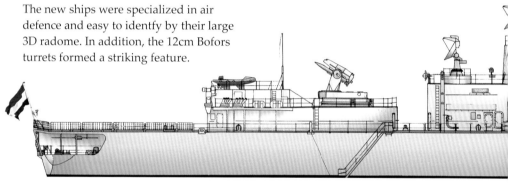

Above: As in port in the seventies.

Below: As being at sea in the nineties.

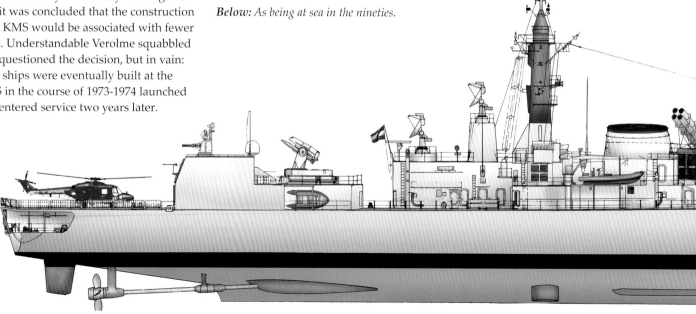

Guided Missile Frigate						
Name	Pennant	Builder	Laid down	Launched	Commis-sioned	Fate
Tromp	F801	KM de Schelde, Vlissingen	4 August 1971	2 June 1973	3 October 1975	Decommissioned 1999. The 12 cm guns preserved by the Naval Museum in Den Helder.
De Ruyter	F806	KM de Schelde, Vlissingen	22 December 1971	9 March 1974	3 June 1976	Decommissioned 2001. Bridge and 3D radome have been preserved by the Naval Museum in Den Helder.

These two 60-ton turrets were removed from the already laid up Type 47 A destroyer 'Gelderland' and were fitted on both GW frigates after overhaul. The ships were equipped with command facilities for a task group commander. Also capable for ASUW (anti surface warfare) and ASW (anti submarine warfare).

Both ships succeeded each other as flagship. Eye catching and fine lined representing 'Holland promotion' world wide.

They took part in many (international) exercises. Special squadron voyages, such as the winter visit to the US in 1982 ('Tromp') and Fairwind'86 to the Far East ('De Ruyter').

After 25 years of loyal service, 'Tromp' was withdrawn in November 1999 and 'De Ruyter' in October 2001.

Their fate was uncertain for a while. The possibility of a transfer to the Indonesian Navy had been suggested. The option for one ship to be prepared as a museum was also considered. But economics rejected these plans, after which it was finally decided to scrap both ships.

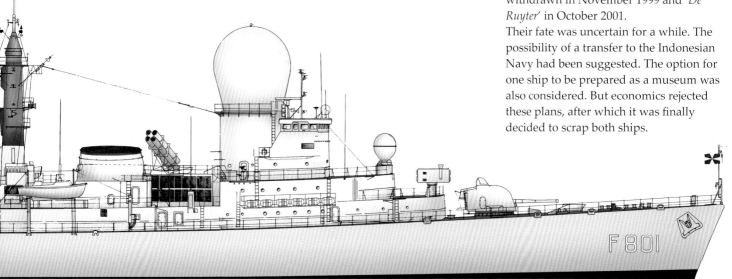

RAL 7038   RAL 7024

In December 2002 the ships were seperately towed out from their homeport. *Tromp* had offered her twin 12 cm Bofors and *De Ruyter* the 3D radome together with her bridge (88 tons). They were placed on the site of the Navy Museum at Den Helder.

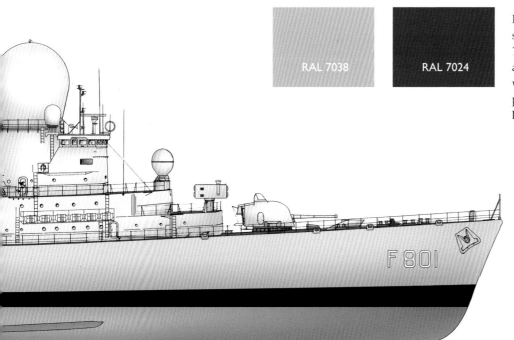

Modelplans

Plans are available at:
1- Netherlands Ministry of Defence: www.defensie.nl/ onderwerpen/ modelbouwtekeningen
2- NVM (Neth. Modellers Association): www.modelbouwtekeningen.nl

1977. Anchored off Torquay (GB).

COMPARABLE NATO SHIPS

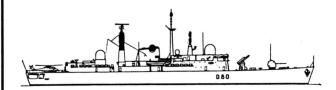

British Type 42 destroyer

The Type 42 or Sheffield class, was a class of fourteen guided missile destroyers for the Royal Navy. The primary role was to provide area air defence. As a radar picket the Type 42 could be stationed 30 nm up-threat. A picket line spread out 15 nm apart. In the Falkland War ('82) a picket line of 3 Type 42's were pushed out to 60nm from the Mainbody (carriers *Hermes* and *Invincible*).

First of class commissioned: 1975 (HMS *Sheffield*)

(See also Warship 09: HMS *Southampton*)

British Type 42 destroyer (batch 1 and 2)	
Displacement	Standard: 3,600 tons, Full load: 4,100 tons
Dimensions	125 m (410 ft) x 14.3 m (47 ft)
Machinery	COGOG; 2 × Rolls-Royce Olympus TM3B high-speed gas turbines, (50,000 shp (37 MW); 2 × Rolls-Royce Tyne RM1C cruise gas turbines, (5,340 shp (3.98 MW)) 2 shafts
Performance	30 knots (56 km/h)
Complement	298
Armament	1 × twin launcher for GWS-30 Sea Dart (24 missiles) 1 × 4.5 in Mk. 8 automatic 2 × 20 mm Phalanx CIWS 2 × 20 mm Oerlikon / BMARC L/70 KBA guns in GAM-B01 single mounts 2 × STWS II triple 12.75 In anti-submarine torpedo tubes

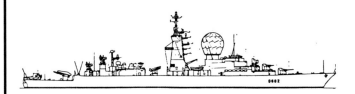

French Suffren class

They were one of the first generation guided missile frigates. Also the first post WWII French warships to be designed from the start as general-purpose vessels.

Visual recognition; Because of the radome the ships resembled Tromp-class. However they seldom operated in same area. France retired from NATO in 1966 to become a full member again in 2009.

First of class commissioned: 1967 (FS *Suffren*)

French Suffren class	
Displacement	Standard: 5,100 - 5335 tons
Dimensions	158 m (518 ft) x 15.5 m (50.9 ft)
Machinery	4 boilers and 4 turbines; 72,500 shp (53 MW); 2 turbo-alternators and 3 diesel-alternators 2 shafts
Performance	34 knots (63 km/h)
Complement	360
Armament	1 x twin launcher of surface-air Masurca missiles (DRBR51-guided) (48 missiles) 4 x MM38 Exocet missiles launcher (4 missiles) 1 x Malafon anti-submarine rocket torpedoes launcher 4 x L5 launcher (10 ASW torpedoes) 2 x 100 mm (3.9 in) turrets 4 × 20 mm guns 4 × 12,7 mm machine guns

OPERATIONAL REQUIREMENTS

For the design, the navy initially cooperated with the Royal Navy to exchange information of 3D radar developments with British info about Dart surface-to-air missiles. After some years this diminished, but later intensified again. Also, when the issue of propulsion was discussed, the British were consulted.

The first draft had geared steam turbines with fixed propeller blades. A proven and safe configuration and ... well known to operate. When the design process developed it changed in favour of gas-turbines. Saving space (no boilers), weight and accelerating readiness. No need to raise steam (gas heats up much faster). Gasturbines are also less noisy, reducing the acoustic signature. The Olympus did deliver high speed power (30 knots), but were not efficient on cruising speeds (18 - 20 knots) and not reversable. This was solved by using controllable pitch propellers. Where the issue of high and cruising/economical speed was solved by doubling the machinery. One turbine for high speed and another for cruising. When high speed was demanded the 'Ollies' were activated.

On 27 July 1970 the order was given to N.V. Koninklijke Mij. De Schelde in Vlissingen (Flushing). The shipyard allotted

The design

The project of design and constructing the GW Frigates was a rupture with the past. For the first time the Royal Neth. Navy was able to follow the path to new ship without foreign help. Making important decisions, solving weight and stabilization issues, and implanting new technologies. Specially the choice for gas turbines was important because it did set the direction for future designs.

As always there was lack of funding and the politics wanted to reduce budget. By the end it turned out that the project did exceed estimates with about 15%.

Wind engineering testing at Netherlands Aerospace Centre (NLR). Note the enlarged engine uptakes,

yard numbers 344 and 345 to be built in a covered dock. Laid down on 4 August 1971 (no 344) and 22 December 1971 (no 345). On 2 June 1973 HM Princess Beatrix baptized yard number 344 and named the ship *Tromp* followed by a floating, since there was no slope. (On 9 March 1974 the ceremony for *De Ruyter* followed).
Fitting out proceeded outside because of the height of the frigates.
They were easy to recognize by their large radome and smaller dishes and domes.

Also remarkable was the absence of portholes and reduced amount of doors. In 1977 and 1978 both frigates had to return to the shipyard because of aluminum cracks in the superstructure. These cracks appeared in the forward deckhouse and in the mid. Aluminum proved less suitable for certain sections of a warship. It could cause hazardous situations for the crew as lessons learned from the Falkands War.

Port visit to Amsterdam.

Building the Guided Missile Frigate

After approval of the design, Koninklijke Maatschappij de Schelde was rewarded with the comstruction order for two frigates. Due to many innovations the costs had increased significantly and while still on the drawingboard the first cuts had already been made by spending less on guns, radio equipment, sonar and monitors for the operations room.

Yard number 344 (Tromp) was laid down on Wednesday 4 August 1971. Building was officially started by Ir. K. de Munter director of the 'Bureau Scheepsbouw' of the R. Neth. Navy by pushing a button initiating the lowering of two sections on the keel and the welding of these.
To meet the agreed timetable and for security reasons a sheltered dock had been built.

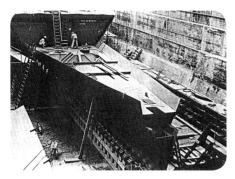

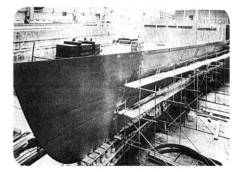

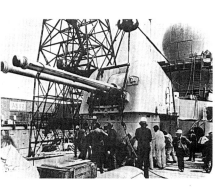

The trials started on Monday 10 March 1975. Visual call signs are not applied yet.

Below: *Starboard yard-arm: hoisted flag Hotel (pilot embarked), to port flying the flag of the shipyard.*

Right:
Photo taken in 1985,
from antenna deck
above the bridge.
At that moment the
ship is preparing for
departure.

Navy Days 1986

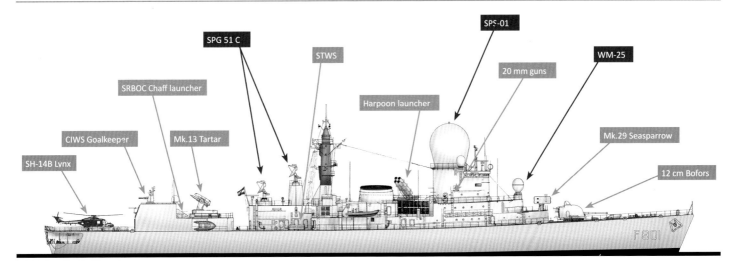

The high freeboard enhanced a dry weatherdeck providing increased safety when carrying out a RAS. Seaworthiness was important because the intended area of operations allocated to the Dutch Navy was the Eastern Atlantic not known for fair weather. To counter the Soviet submarine threat NATO wanted to control the GIUK-gap, the ocean areas between Greenland, Iceland and the UK.

Lessons Learned

High freeboard, raised forecastle deck. Experiences of convoy escorts in the Atlantic, the "submarine hunters of WW2" led to the designs of ships with a raised forecastle deck and a high freeboard like the British Type 12's and Leanders.

Portside bridge deck.

Crane at the rear side of the hangar.

Den Helder 15 May 1975. Zuiderkruis on trials entering the "Nieuwe Haven" for the first time to meet Tromp, *jetty 20- also on trials-; both ships were not yet commissioned. Dutch Leander type* Van Speijk *berthed alongside jetty 21.*

The name "Tromp" is a remembrance dedicated to two famous admirals.

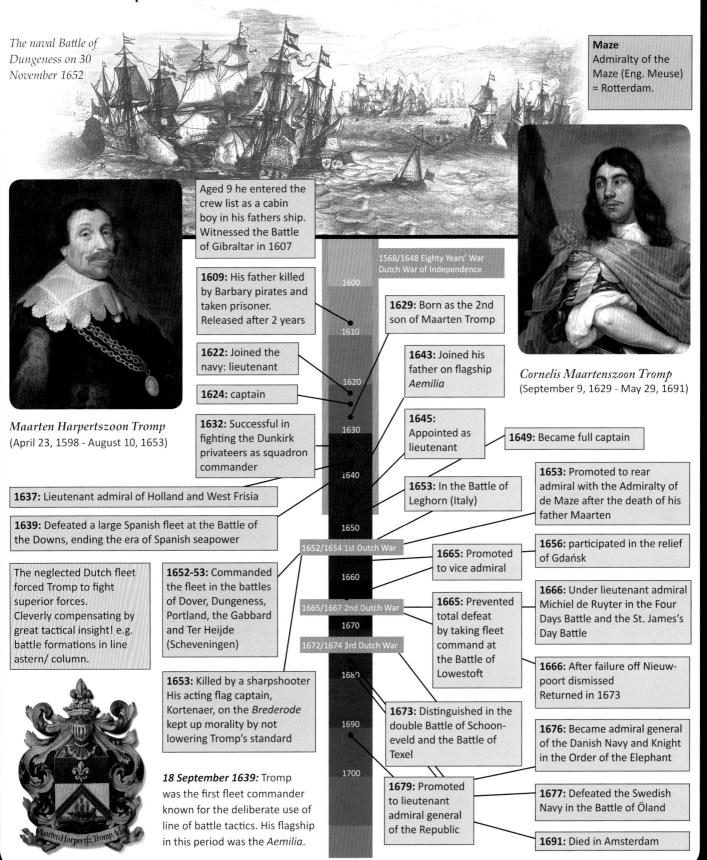

The naval Battle of Dungeness on 30 November 1652

Maze
Admiralty of the Maze (Eng. Meuse) = Rotterdam.

Aged 9 he entered the crew list as a cabin boy in his fathers ship. Witnessed the Battle of Gibraltar in 1607

1568/1648 Eighty Years' War Dutch War of Independence

1609: His father killed by Barbary pirates and taken prisoner. Released after 2 years

1629: Born as the 2nd son of Maarten Tromp

1622: Joined the navy: lieutenant

1643: Joined his father on flagship *Aemilia*

1624: captain

Cornelis Maartenszoon Tromp (September 9, 1629 - May 29, 1691)

1632: Successful in fighting the Dunkirk privateers as squadron commander

1645: Appointed as lieutenant

1649: Became full captain

Maarten Harpertszoon Tromp (April 23, 1598 - August 10, 1653)

1637: Lieutenant admiral of Holland and West Frisia

1653: In the Battle of Leghorn (Italy)

1653: Promoted to rear admiral with the Admiralty of de Maze after the death of his father Maarten

1639: Defeated a large Spanish fleet at the Battle of the Downs, ending the era of Spanish seapower

1656: participated in the relief of Gdańsk

The neglected Dutch fleet forced Tromp to fight superior forces. Cleverly compensating by great tactical insight! e.g. battle formations in line astern/ column.

1652-53: Commanded the fleet in the battles of Dover, Dungeness, Portland, the Gabbard and Ter Heijde (Scheveningen)

1652/1654 1st Dutch War

1665: Promoted to vice admiral

1665: Prevented total defeat by taking fleet command at the Battle of Lowestoft

1666: Under lieutenant admiral Michiel de Ruyter in the Four Days Battle and the St. James's Day Battle

1665/1667 2nd Dutch War

1672/1674 3rd Dutch War

1666: After failure off Nieuwpoort dismissed Returned in 1673

1653: Killed by a sharpshooter His acting flag captain, Kortenaer, on the *Brederode* kept up morality by not lowering Tromp's standard

1673: Distinguished in the double Battle of Schooneveld and the Battle of Texel

1676: Became admiral general of the Danish Navy and Knight in the Order of the Elephant

18 September 1639: Tromp was the first fleet commander known for the deliberate use of line of battle tactics. His flagship in this period was the *Aemilia*.

1679: Promoted to lieutenant admiral general of the Republic

1677: Defeated the Swedish Navy in the Battle of Öland

1691: Died in Amsterdam

Maarten Harpertsz Tromp Adm.

1600
1610
1620
1630
1640
1650
1660
1670
1680
1690
1700

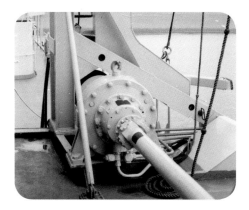

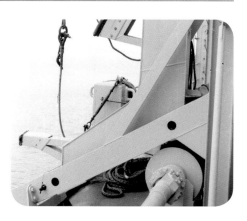

The headline of the ad reads: 'Navy frigate HNLMS Tromp *visiting the big city', over the years the flagship visited New York three times. This type of illustrations often appeared in recruiting advertisements. Join the navy, see the world!*

Right:
Mainmast with helicopter glide path indicator.

Bottom:
The white painted area on deck indicates storage of ammunition below.

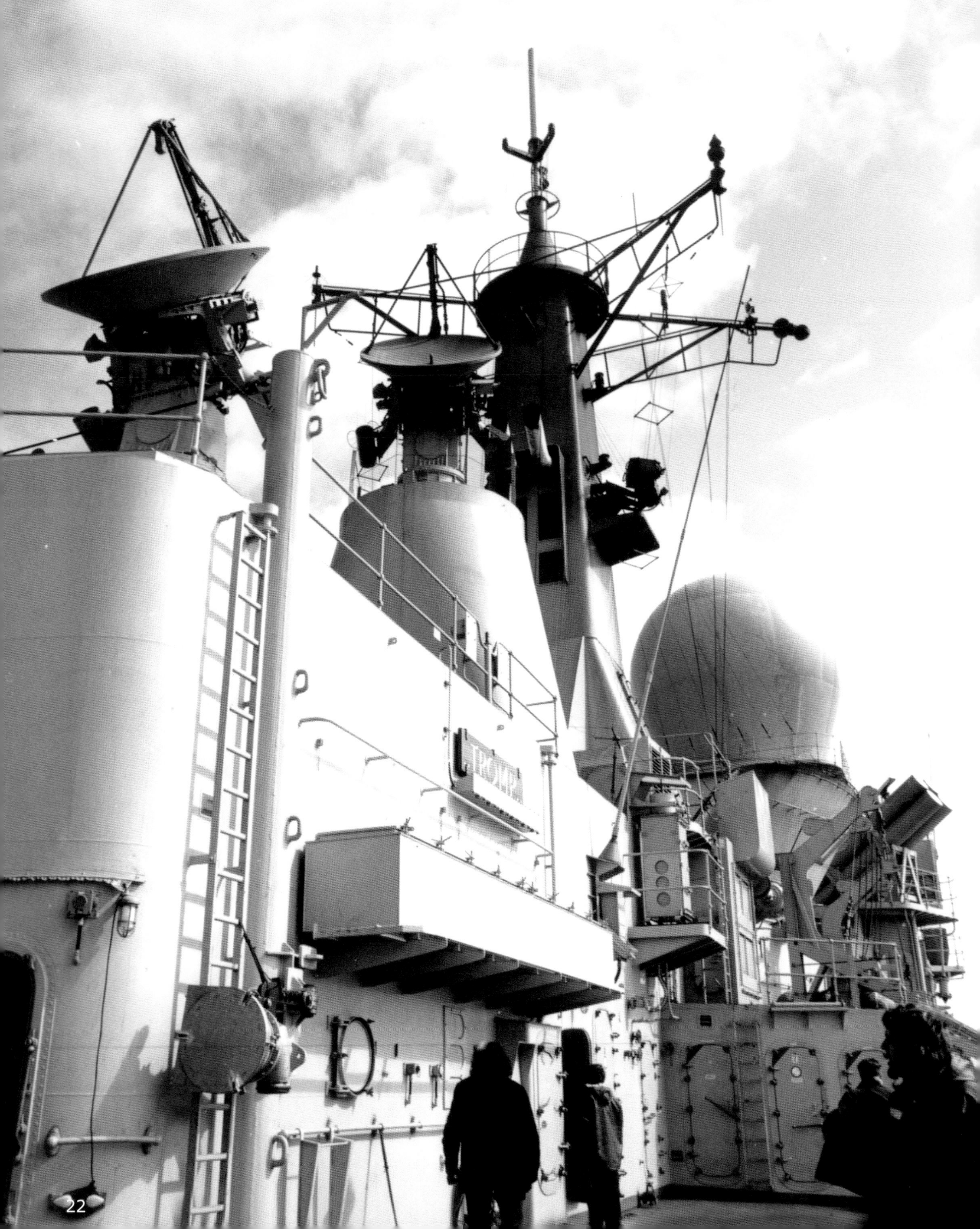
TROMP

SENSORS AND ELECTRONICS

3D radar – SPS-01

The most important and striking sensor was the 3D-radar that gave the ships their easy to recognize silhouette. It had a diameter of 9 meters. A constant temperature was kept inside to avoid interference by outside weather conditions. Resulting in a high energy demand in arctic and tropical areas. The dome was carried by a construction that was built against the bridge. This radar had a range of 250 miles and was capable of handling simultaneously over one hundred aircraft tracks. Collected data was automatically forwarded to the computer system for processing and analyzing were it instantly generated a list of the six most dangerous threats to counter with Tartar.

The air-search functions were performed by a pair of parabolic reflectors mounted back-to-back, and the tracking functions by a similarly mounted pair of planar, phased-array antennae with integrated IFF. These antennae where mounted on a single platform with common turning gear and were housed within a large fibre-glass dome. The back-to-back arrangement gave a very high data-rate, and high and low cover was provided by the system. Initially there were many problems with the radome. It was constructed by plastic

Preparing to install the 3D radar on De Ruyter. (NIMH)

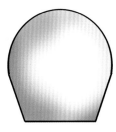 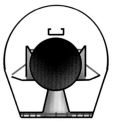 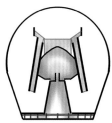

panels who turned out to be too weak to repel long term weather conditions. After research, a new radome was designed.

Using another construction and slightly changed model. On the photographs differences can be seen clearly.

  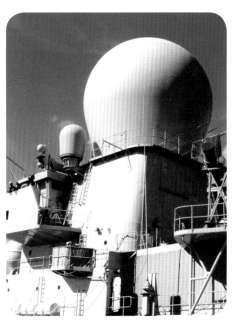

*The weather punished the radome in such a hard way that it needed to be replaced several times. **Left:** Original dome 1975 like patchwork.*
*__Centre:__ In 1980 an improved model was mounted. **Right:** In later years very smooth like mr. Kojak's skull...*

Initially the radar was called the MTTR (Multi Target Tracking Radar) later this was changed in SPS-01[*]. It combined search and target tracking (air warning and target designation) in a single multiple antenna. SPS-01 was designed to meet the mutually contradictory requirements of search and multiple-target tracking. For search, it had to radiate over a very wide volume, with a relatively low data rate for long ranges and a higher one for short range; accuracy could be sacrificed for a high probability of target-detection. For tracking, on the other hand, precision was necessary, coupled with a concentration of radiated energy in the direction of the target, and a very high (or, better, continuous) data rate. Discussions with the Royal Neth. Navy began in 1958, the goal being simultaneous high-precision tracking of multiple targets; SPS-01 was now credited with the capability to handle over a hundred air targets.

Research commenced in 1959, and a prototype was available in February 1964, followed by the production of two systems. Testsing started in 1967 and were completed in 1969. The MTTR project was for some time a joint project with the Royal Navy, and its variant (Type 988, or 'Broomstick') was intended for the Type 82 missile destroyer (*Bristol*) and for the abortive British carrier project (CVA-01).

The radar employed six antennas, all mounted together on a stabilized base and rotating together at 20 rpm: two paraboloids back-to-back, with feedhorns moving to generate any of five alternative beams plus a sixth (fixed) low-angle beam; two

back-to-back FRESCAN[**] arrays at right angles to the parabolas; a multi-element antenna for high-angle search (omitted in production models); and a slotted waveguide for IFF mounted below one of the parabolas. In operation, the low-angle search beam is operated continuously to provide short-range warning with a high

data rate (40 scans per minute). The other five beams were energized in sequence one by one during successive rotations, so that for each one the data rate was reduced to 8 per minute. The net search pattern was a cosecant-squared shape but avoided the drawbacks of a more conventional fan-beam radar, i.e. low antenna gain and great susceptibility to clutter and jamming. Beam dimensions varied with elevation:

---

[*]   SPS - Signal Processing System

[**]   FRESCAN is acronym: Frequency-Scanning

in each case the paraboloid assured a horizontal width of 1.5° but the vertical beam size varied between 2° at an elevation of 2.5° to 30° at an elevation of 21°. Targets were detected automatically, and their two-dimensional coordinates fed into the central computer, which instructed the two FRESCANs to end them in three dimensions for tracking.

Each of the FRESCANs consisted of a series of slotted waveguides slanted to the horizontal and fed by the usual sinuous waveguide for vertical scanning. The slant made it possible for the system to perform a true vertical scan while the antenna rotated rapidly; in effect the FRESCAN was used to scan both vertically and horizontally, and thus to obtain precision

target data. Approximate beam dimensions were 1.5° x 1.5°, the latter depending on beam elevation. Back-to-back positioning provided a data rate of 40 scans per minute. In operation, the computer first ascertained that a target detected in two dimensions was not already being tracked. It then programmed a frequency sweep which measured target elevation; on the next scan the frequency variation was programmed to give a horizontal scan for precision target-bearing location. Each target was scanned crosswise at the rate of twenty measurements per minute. In this way a pure FRESCAN was used for true three-dimensional tracking, without phase shifters; however, its effective tracking rate was limited by the requirement for mechanical rotation.

1988. First in the navy Tromp *received two small domes for the NATO-SATCOM III system (satellite communication)*

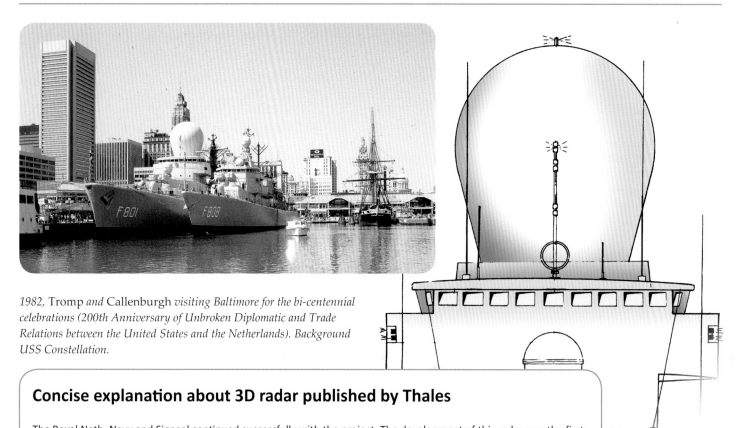

1982, Tromp *and* Callenburgh *visiting Baltimore for the bi-centennial celebrations (200th Anniversary of Unbroken Diplomatic and Trade Relations between the United States and the Netherlands). Background USS Constellation.*

Concise explanation about 3D radar published by Thales

The Royal Neth. Navy and Signaal continued successfully with the project. The development of this radar was the first effective strategy to counteract a hostile military threat at sea. The basic architecture of the rotating antenna system consists of two parabolic search aerials (back-to-back) and two planar aerials (back-to-back) for the tracking function. Each parabolic search aerial has a variable feedhorn to set the elevation of the search beam. The planar tracking aerial consists of a number of slotted waveguides which are end-fed by the rf-source. Each transmitter pulse has a computer controlled frequency. During rotation either a pure vertical scan (for height finding) or a cross-shaped scan (for 3D position measuring) can be performed by proper programming the frequency of the successive rf transmitter pulses. This rotating 3D radar represented the techniques, which were available at that time.

The new 3D radar produced such a large data flow that automation of these data was inevitable.

The fast development of digital computerisation and micro-electronics made it possible to lead these flows via interfaces direct into the successive tactical processes, such as threat evaluation - target selection - weapon selection, without human interference, the so called DAISY.

Below:
Flag in 'sea position'; without Goalkeeper and Satcom.

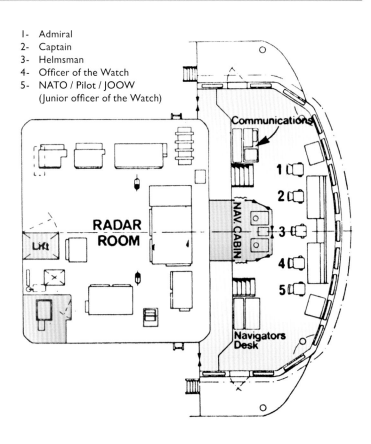

1- Admiral
2- Captain
3- Helmsman
4- Officer of the Watch
5- NATO / Pilot / JOOW
    (Junior officer of the Watch)

Note the small porthole in front of the radome

Below: *Compare with page 26: fully armed with Harpoons and equipped with Goalkeeper and Satcom.*

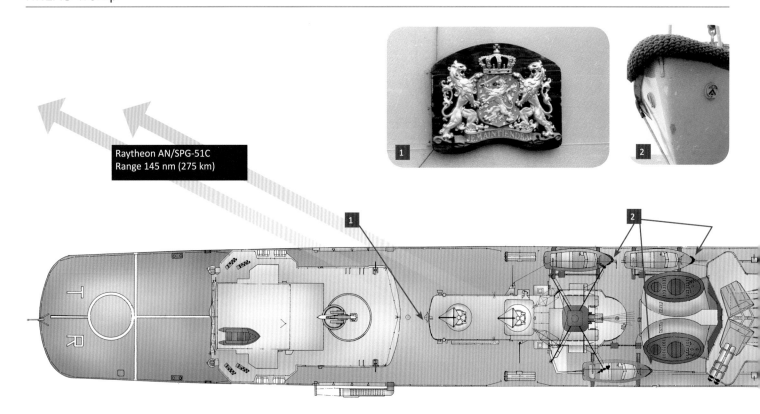

Raytheon AN/SPG-51C
Range 145 nm (275 km)

Weapons engineering specialists carry out routine maintenance at WM-25 and Sea Sparrow launcher. The WM-25 system could engage one air target at a time with Sea Sparrow missiles out to about 10nm. That means that at any given time the Tromp could simultaneously engage a max of 3 surface-to-air missiles (1 Sea Sparrow, 2 Standard) in the air against up to 3 targets.

WM-25

About 1970 Signaal developed the first generation of this type. The waterproof egg saved weight by eliminating the requirement to waterproof antennas separately, dry air was pumped directly inside. Because both antennas used the same transmitter there was no need to shift range scales between them. The result was a very rapid reaction. The chief disadvantages of this were a relatively low location for the search antenna and possible blind spots.

The WM-25 could control a semi active surface-to-air missile and had 2-gun channels. It could simultaneously track one air, one surface and one shore target: or one air and two surface targets. The egg-shaped dome (diameter 2.39 m, height 3.26 m. weight 780 kg) enclosed a search antenna below and a tracking dish above. During search all power was fed to the lower antenna, but once a target had been detected. Some power was fed into the tracker. Signaal claimed the mounting of both antennas on a single platform climi nated errors in transferring a target from search to track. The search antenna could also be used to designate to a separate tracking antenna STIR (Signaal's Tracking and Illumination Radar).

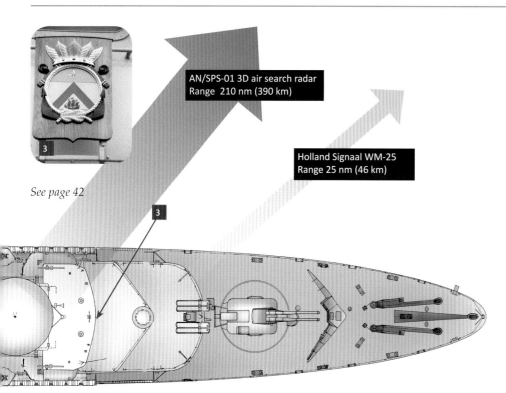

AN/SPS-01 3D air search radar
Range 210 nm (390 km)

Holland Signaal WM-25
Range 25 nm (46 km)

See page 42

The portside Decca radar antenna on bridge roof, was later relocated.

Decca radar

For navigation purposes and helicopter control the ship was equipped with Decca dual Transar radar for which twin port and starboard antennae above the bridge, provide 360° coverage. The system was selected because of clutter problems posed by the siting of the 3-D radome. ECM sensors are carried on the mainmast and are backed up by two British-design Corvus chaff dispensers abreast the hangar.

For service in tropical waters the roof of the bridge was painted white.

Mainmast in 1986 and 1999 (R).

At the base of the mainmast the emergency control position. (see page 31)

Early picture of Tromp. *Flying the NATO flag. Note; only one canister launcher for Harpoon.*

Between base of mast and flag lockers.

9 December 1977. Tromp, show-piece of the navy was proudly presented to act as flagship in the STANAVFORLANT. Tromp departed Den Helder 4 Januari 1978, staff embarked. Bare mast; below the white pole of the URC-246 UHF aerial only the "masthead obstruction lights" (compare with p. 30).

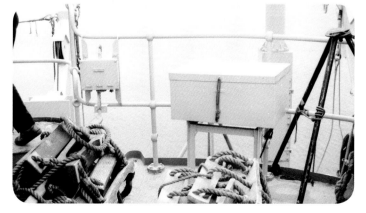

Netherlands in NATO

The Dutch were proud (then...) to be considered the most principal smaller Navy in NATO.

Since the formation of STANAVFORLANT in 1968 it was in 1977/78 the third time a Dutch Flag Officer was appointed in charge of this NATO squadron. Commodore K.H.L. Gerretse embarked in *Tromp*. This new FFG with a strong AAW capability was equipped with an advanced Action Information System and a 3D (three dimensional) radar. The Commodore was an experienced submariner.

Saluting battery

On Tartar deck some small caliber special applied guns mounted for ceremonial purposes. The number of rounds fired in a salute varied depending on the conditions. Circumstances affecting these variations included the particular occasion of the salute or celebrations, the branch of service, and rank (or office) of the person to whom honours are being rendered.

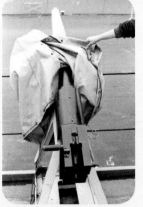

Sonar

ASW data was provided by two hull-mounted sonars. One was the long-range EDO(USA)CWE 610, a low-frequency sonar in production since 1969 which was also in service with the Italian Navy. The other was the even more recent Type 162M developed for the Royal Navy. This had a sideways-looking scan pattern and was designed to classify medium depth and seabed targets out to about 1200 yards.

Sonar dome of Tromp *sustained some damage.*

DAISY

Information gained by the sensors was transmitted to the Tactical Data Handling System or in Dutch: DAISY (Digitaal Automatisch Informatieverwerkend Systeem) a data handling system to shorten the response time assisting in selecting and activating weapon systems. It was not fully automatic, so the human factor was decisive during operation. Once the computer had processed and saved the information it was available on every screen by selecting the right button. The radar-operators sitting in font of their desk had now access to the required information for their specific task, like air or surface. Plotting info once written on bulkhead-mounted perspex toteboards was now available by a close-circuit TV system. So every operator could access video images on his monitor (for instance to monitor flight operations in the hangar/helideck) The operator had a "split head set" which enabled him to communicate internal and external.

All weapons and sensors were linked by a command system designated SEWACO-2 (in Netherlands language: **SE**nsoren, **WA**pen- and **CO**mmandosystemen). The electronics for this were mainly of Dutch manufacture supplemented by components from other NATO countries. The computer-based command system processed information from all sensors, evaluated threats and designated targets to the weapon systems.

The six most urgent threats were automatically ranked and presented to the operator. The operations room, in which housed the display sub-system, was located low in the ship. The sonar room was in a corner behind a sliding door, adjacent to computer and conference rooms.

Equipment for electronic warfare was also carried. DATALINK was used for exchange digital - tactical information with other ships of the Task Group.

Quarterdeck

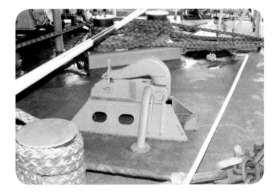

MACHINERY

One of the turbines of Tromp

In the original design the frigate was driven by conventional steam turbines with fixed pitch propellers. At that moment the Royal Navy was developing a new generation of aero-derived gas turbines to be installed in their designs. The major attraction of gas turbines for the new frigates were the reduction of engine room personnel needed to operate and maintain them and the low noise signature compared with steam turbines and diesels. Making them much better suited to anti-submarine operations. The frigates finally emerged from the drawing board with basically the same COGOG system (**CO**mbined **G**as **O**r **G**as) like the British Type 42 (Warship 9) destroyers. The British Y-ARD (Glasgow) acted as consultant for the main machinery installation and controls. Where both countries entered a joint agreement to monitor the performance of the turbines.

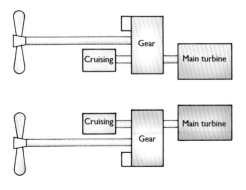

Gas turbine

Ending 19th century the idea of a gas turbine propulsion appeared. Early 20th century a first installation was tested but proved not very efficient. The first successful gas turbine was build in 1939 by the Swiss Brown Boveri. After WWII the development of the gas turbine for ship propulsion really took off. In the years 1960-1970 several navies built new ships powered by gas turbines. Some others in combination with a diesel or steam powerplant for propulsion.

The advantages of gas turbines:
- Short periods of maintenance (replace and repair)
- Fast start up time
- Short response when manoeuvering
- Suitable for remote control

The main gas turbine or the cruising turbine was driving the gearbox and propeller shaft. The frigate had two shafts with controllable pitch propellers. The two main turbines were the Rolls Royce Olympus TM 3B. Each of 22,000 hp and located in the forward engine room. These were down-rated turbines of the 27,000 hp version in Type 42 destroyers to extend the life of the gas-generators. Thereby reducing maintenance costs.

The two Rolls Royce Tyne RM 1A of 4,000 hp cruising turbines were installed in the after engine room. Also, the by Schelde manufactured gearboxes were located in this room.

Gas turbines were connected to the gearboxes by so-called S.S.S. links (Self Shifting Synchronizing). When for instance the cruising turbine was driving the shaft the main turbine could be started and when the rpm's increased the cruising turbine gearbox automatically shifted

drive. And reverse from main to cruising turbine when the rpm's decreased. Propulsion was controlled by Netherlands designed Fokker remote system. It was possible to control the engines from the technical center, operations room or bridge. Using D.T.S. (Drukknop Telegraaf Systeem (Push button telegraph system)) automatically the correct rpm and propeller blade angle was set up. In case of malfunctioning, the turbines could be operated locally. Therefore, an emergency manoeuvring panel was installed in the aft engine room.

Important machinery data was monitored by the DECCA-ISIS system and presented in the technical center. Around 400 points were continuous measured, when a malfunction appeared an alert was presented to the duty engineer. Therefore, the engine rooms were most of the time unmanned.

ELECTRICAL INSTALLATION

A brand-new Tromp in US waters to collect the SSM Harpoon.

The main power supply was of 4 diesel generators of 1000 kW each. Diesel rather than gas generators were selected partly because of the problems of providing additional air intakes and uptakes to those required by the propulsion machinery but also because of the high fuel consumption of gas turbines at low ratings.

Two were installed in the forward engine room, and two in the aft engine room. It was possible to connect these generators from the main switch board. The diesel generators could start up and shut down automatically when the demand urged them. It was controlled by the ADS (Automatisch Diesel Startsysteem). Also, the electrical system was controlled by a panel in the technical center.

The exhaust of the diesel generators was installed in the main mast.

The arrangement of propulsion and auxiliary machinery was designed to minimize action damage. Either set of turbine could continue to operate with the other engine room flooded and the gear boxes were designed as watertight units. Furthermore, the separation of the two pairs of gearboxes by watertight bulkheads lessened the likelihood of power failure. All machinery was designed for the quietest possible operation: the diesel generators were fully enclosed in noise-absorbing hoods, the cruise turbines resiliently mounted, the gearboxes made exceptionally quiet by silencers installed in the intakes and uptakes.

Left:
Passage to the heli deck.

Right:
Port side superstructure in center.

WEAPON SYSTEMS

The armament finally fitted is both comprehensive and well-balanced, with two weapon systems provided for each of the three functions - surface-to-air, surface-to-surface and ASW - demanded by Staff Requirements.

Tartar

The main surface-to-air element is the American Standard SM-1 missile (of which 40 are carried). At first the British Sea Dart was considered but compared to Tartar

Tromp *employed a Mk 13 trainable single-rail mechanical missile launcher with a magazine containing SM-1 Standard-surface-to-air missiles.*
The Mk 13 could fire 1 missile every 8 seconds.

Right: To intercept air targets at medium range (out to about 25 nm) SM-1 Standard (MR) would be launched receiving continuous info by a SPG-51C tracker-illuminator which could deal only with one target at a time.

Left: Tartar/SM-1

this was to be too expensive and heavy. The single-arm Tartar Mk 13 launcher is mounted on a deckhouse which extends forward of the helicopter hangar, giving good all-round coverage supported by two AN/SPG-51C fire control radars aft of the mainmast. Tartar/SM-1 has been adopted by NATO and other navies. This installation differs from all others, however, in that target-tracking information for the SPG-51s was initially provided not by a radar of American manufacture, such as the SPS-52 and -48A planar radars, but by the massive HSA three-dimensional radar specifically designed for these ships.

Nato Sea Sparrow

Back-up for Standard was provided by the NATO Sea Sparrow short-range surface-to-air missile. Developed from the American Basic Point Defense Missile System (BPDMS), it did use a special version of the Sparrow missile which has folding wings and could therefore be launched from a new lightweight 8-cell launcher (its predecessor used a modified ASROC launcher). With an effective range of about 10 nm

the missile could engage targets (sea skimmers) down to a level of about 8 m above the surface. The Sea Sparrow missile was coupled with the WM-25, which also served the 120 mm mounting. The WM-25 was the latest in the popular M20 series of radars which now equip many of the world's small warships and which, like the 3-D radar, are manufactured by Hollandse Signaal Apparaten.

Tromp also employed a manually reloaded Mk 29 8-round launcher for RIM-7M NATO Sea Sparrow missiles.

It comprised a tracking/illumination radar above a parabolic search reflector on a common mount, housed in a near-spherical radome. The tracker radar directs the Sea Sparrow missiles. The search radar provides air and surface search as well as automatic tracking for one or two surface targets. The whole system had been designed for quick reaction against medium- and close-range targets, and computer-direction ensures an all-weather capability.

12 cm Bofors

Tromp carried two Swedish Bofors 120 mm (4.7") automatic dual purpose guns against air and surface targets in a twin mounting. When introduced in 1950 it was considered "a most advanced gun" by the US Bureau of Ordnance, "The best European naval gun available". In the 1950's 24 twin mounts were purchased for the ASW-destroyers and proved to be very reliable after elaborate tests and trials. While still on the drawing board it was decided to transfer the mountings of destroyer *Gelderland* (1955-1974) to the new frigates.

It was saving money in the ever increasing expenses of the project. The underside of the mountings were modified to fit the new frigates

Technical data

	Bofors M1950
Designation	Sweden: 12 cm/50 (4.7") Model 1950 Netherlands: 120 mm Mk 10
Date In Service	1950
Calibre	120 mm
Rate of fire	42 - 45 rounds per minute (42 gives the optimum dispersion)
Max Range with 52 lbs. (23.5 kg) HE	45 degrees: 20,890 yards (19,100 m) Ceiling: about 29,800 feet (9,000 m)
Train rate	25 degrees per second
Elevation	-10 / +85 degrees 40 degrees per second
Train	about +150 / -150 degrees
Total weight	148,000 lbs. (67,000 kg)

The Nato Sea Sparrow and Bofors twin 120mm rapid-fire gun were controlled by Signaal WM-25 fire control system covering the spot forward where the SPG-51 radar's were blind. Sea Sparrow was more effective against low-altitude targets than Tartar.

Planned Ikara

The Australian designed Ikara long range A/S weapon was not selected.
MATCH, a more flexible multitask system was preferred. Acronym stands for:
MAnned **T**orpedo **C**arrying **H**elicopter.
NATO: Medium-Range Anti-Submarine Torpedo-Carrying Helicopter.

Note:
If the helicopter had to scramble the order given was: "Action MATCH".

STWS

Defence against submarines at short range Mk. 32 triple torpedo tubes were mounted. The STWS (**S**hips **T**orpedo **W**eapon

System) offered the possibility to attack submarines in shallow waters. Firing Mk.46 Mod.5 torpedoes, range 5,9 nm (11 km) at 40 kts.

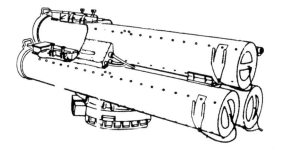

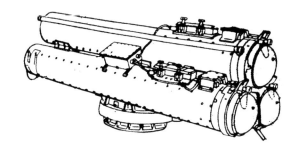

Harpoon

It took some time to choose a surface-to-surface missile to complement the twin Bofors. The main contenders were the French MM38 (Exocet) and the American Harpoon. Eventually a decision was made in favour of the more advanced (and more costly) Harpoon, for which ramps can be seen just forward of the twin uptakes.

homing radar, which is frequency-agile to prevent jamming, then select and locks-on the target. In the final phase the missile executes a climb /dive manoeuver. At a length of 4.63m (shorter than the horizon-range Exocet) Harpoon was particularly compact for a SS-missile. Although Harpoons also could be launched by the Tartar launcher, the ship had weather-

Technical data	
	Harpoon
Designation	RGM-84
Date In Service	1977
Class	Subsonic Cruise Missile
Length	4.628 m
Diameter	0.343 m
Guidance	Inertial, semi-active radar
Speed	0.85 Mach (High subsonic)
Range	67 nm = 125 km
Weight	681 kg (224 kg warhead)

In 1971 McDonell Douglas was selected by the US Navy as prime contractor for developing the missile. By August 1996, 6265 missiles were delivered to USN and foreign customers.

The missile is a 'fire and forget' weapon, being provided with pre-launch target data, using inertial guidance techniques during its outward flight and finally its own radar seeker for terminal guidance. Making the missile independent of the ship's sensors. It cruises at low altitude directed by altimeter control. The active radar homing system switches on automatically at a predetermined distance from the position of the target at launch. The

proof canisters bolted to each of the two supports for a total of eight missiles.

The canisters were in 'shock-resistant' supports mounted such that the missile efflux is directed in a deck mounted deflector shield. (A simple open box structure with outward sloping sides) The Harpoon canisters were mounted high on the superstructure between the twin funnel and radar dome.
Most of the time only two canisters at each side were mounted. (Brown cover for exercises. Red for carrying warhead).

Chaff

Against missile threats the Corvus D Chaff launcher was mounted. This system comprised two multi-barrelled rocket launchers, a control and firing panel and Chaff (radar counter-measure) rockets.

The Harpoon RGM-84C, over-the-horizon, anti-ship missile was carried in canisters,

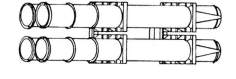

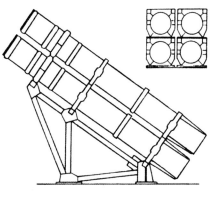

Following launch and dispersion, the system was designed to lure hostile missiles away by creating false targets. {C = Confusion / D = Distraction} Later, in the eighties this system was replaced by the BAE Systems Mk 36 SRBOC (Super Rapid Bloom Offboard Chaff), a deck-mounted, 6-barreled mortar-type array. (pictured left).

Wasp lashed on deck. Although designed to carry the Lynx which entered service in 1976 the frigate often embarked the older, smaller Wasp helicopter. The initial 12 ASW helo's were delivered in 1966/67 (One replacement in 1974) to operate from the Van Speijk class. Over the years three had been written off. The remaining ten transferred to the Indonesian Navy in 1981.

WESTLAND HELICOPTERS
ASW = ANTI-SUBMARINE-WARFARE

The principal anti-submarine weapon carrier was the Westland Lynx which entered service in 1976. This new helicopter had over twice the weight of the Wasp. In its ASW role the Lynx carried two American Mk 46 active/passive acoustic homing torpedoes. These could also be launched by the torpedo tubes on the weather deck.

Within the navy the 24 Lynx helicopters flew in their 36 years of service (1976 - 2012) more than 160,000 hours. The Lynx helicopter, or officially Westland WG.13, was first flown on 21 March 1971 and six prototypes were ordered, followed by seven pre-production prototypes to speed up development. Service trials began in 1976 with No. 700L Naval Air Squadron at RNAS Yeovilton, Somerset. This was an Anglo-Dutch operational evaluation unit. Almost at the same time, the first six Dutch airframes were produced. The Mk. 25 model was called UH-14A in Dutch service. Initial buy was for six SAR helicopters, followed two years later by ten ASW-versions (Mk. 27) called SH-14B. A third batch was for eight MAD-equipped SH-14C's (Mk. 81), delivered back to back to the 'B'-models. By the summer of 1980, all 24 helicopters had been delivered. These models compared roughly to the Royal Navy HAS Mk.2 version.

A standardization program called STAMOL (Standardization and Modernization Lynx) in the early nineties brought all versions up to the same specification and the 22 remaining[*] airframes ended their useful lives as SH-14D models. Some life extension programs had to be performed between 2004 and 2010 to keep ten flying, necessitated by the repeated delays of its successor, the NH-90.

In the early 80s an additional 8 helicopters were ordered. (Reg. 276 up to and including 283) These Westland Lynx Mk. 81 were called SH-14C's and intended for ASW duties.

[*]   In Nov. 1998 at De Kooy: SH-14D (282) crashed during take-off.

The Ten Westland Lynx Mk.27's were delivered 1978/79 and referred to as UH-14B. (Reg.No.266 t/m 275) Pictured is 272 after modifications.

Stationed at NAS De Kooy, near Den Helder the helicopters were operated by the resident squadrons, nos. 7 and 860, the former being the training and SAR-unit, the latter the shipbased ASW and special missions unit. A hydraulic winch (hoist) could lift 600 Lb (270 kg) was an important asset to rescue shipwrecked sailors. Underslung loads up to 3000 Lb (1360 kg) could be carried on an external freight hook.

Squadron 7

- Since 1962 designated SAR squadron
- In 1974 based NAS De Kooy situated on 2½ nm from the North Sea
- In 1978 equipped with Lynx
- In 2006 (Oct) the 1200th rescue mission performed by a Lynx team

When deployed on a ship, they received a 'war-cry' of a cartoon figure: *Gaston*, *Pink Panther*, *Turtle Flight* (Ninja Turtles), *Obelix*, (Bugs) *Bunny*, *Speedy* (Conzales), *Simpson, Dalton* (Brothers), *Snoopy, Buldog, Ducks United* and *Harry*. These were put on the nose and later due to radar signal disturbance on the cabin door. The names were not to 'stick' to a specific helicopter; after the tour all the decorations were removed. For instance; for 'Turtle Flight' ten different registrations were noted throughout the years.

The white markings on the flight deck changed in the eighties

Left: *Lynx helicopter of frigate Van Speijk named 'Road Runner' transferring personnel.*

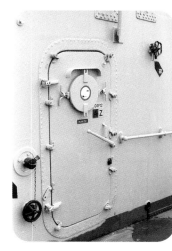

Above, right: *Bulkhead of super-structure forward.*

Right: *Cruising off the English coast.*

Bottom: Tromp *leading frigates* Van Speijk *(1967-1985) and* Van Kinsbergen *(1980-1995).*

Accomodation

Left and right: The hangar was cramped. Besides a helicopter lots of other items were stored. Even some private properties like surfboards and bicycles.

Playing cards in the Wardroom.

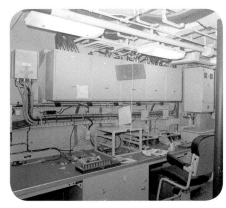

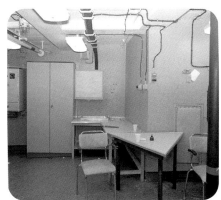

Impression of the working areas in the frigate.

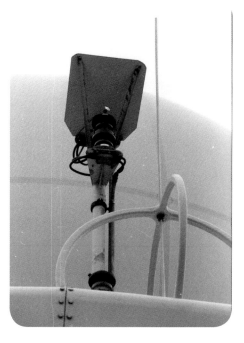

The NLTG about 1987:
De Ruyter *(F 806),* Zuiderkruis *(A 832),* Tromp *(F 801),*
Pieter Florisz *(F 826) and* Jan
van Brakel *(F 825).*

Left:
Flight path indicator.

Right, center and below:
Tartar deck.

HNLMS Tromp

3 October 1975 - 12 November 1999

Commissioned 3 October 1975	First foreign port call 24 November 1975 Wilhelmshaven	Port visits 88 different foreign cities	Crew over the years About 4000 military
Named *Tromp* 8th ship	First time Suez Canal 31 March 1979	Most often visited port Lisbon (10 times)	Ships log 510,800 nm (*De Ruyter*: 620,200 nm)
First exercise 11 April 1975 with HMS *Sheffield*	Most distant Port Perth (Fremantle) 9800 nm (18,150 Km)	Last foreign port call October 1999 Southampton	Fate Scrapped

Smoking was common. As a PR gift a match booklet of the ship was much appreciated!

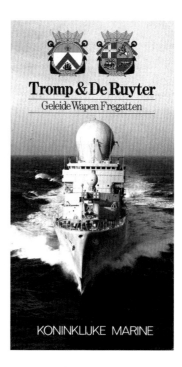

Commanding Officers

From	To	
1975	1976	Ktz J.D. Backer
1976	1977	Ktz J.H.B. Hulshof
1977	1978	Ktz J.H. Scheuer
1978	1979	Ktz P.A.A.J. van Oppen
1979	1981	Ktz R. den Boeft
1981	1982	Ktz J.D.W. van Renesse
1982	1982	KLtz P.K. Kleimeer (tmp)
1982	1983	KLtz J.P. de Klerk (tmp)
1983	1985	Ktz H.A.J. Nijenhuis
1985	1986	Ktz C.P. Klavert
1986	1987	Ktz A. van de Sande
1987	1988	Kltz R.A. Baljeu
1988	1989	Ktz H.W. van Vliet
1989	1990	Ktz J. Waltmann
1990	1991	Ktz C. van Duyvendijk
1991	1992	Ktz A. Vos
1992	1993	Ktz P.C. van der Graaf
1993	1994	Kltz A. Verbeek (tmp)
1994	1995	Ktz J. van der Aa
1995	1996	Ktz J.W. Kelder
1996	1998	Ktz M.A. van Maanen
1998	1999	Kltz S. van der Sluis

The badge of Tromp was derived from the crest of Maarten Harpertszoon Tromp. Granted by the French king Louis XIII when he made him a knight in the Order of Saint Michael (French: Ordre de Saint-Michel). After defeating the Spanish armada in 1639 off the Downs.

Info booklet for guests released by Public Relations.

Above: At the Float Out Ceremony, H.R.H. princess Beatrix named Tromp. Grandmother Queen Wilhelmina launched flotilla-leader Tromp 36 years earlier. The pre-WW2 Tromp had about the same dimensions and displacement.

Zaterdag 2 juni 1973

binnenland buitenland

Prinses Beatrix doopt de opvolger van een „gelukkig" schip

The newspaper reads: *Princess Beatrix christened the descendant of a 'happy' ship, the seventh* Tromp. *Referring to the light cruiser* Tromp (Warship 01), *called 'The Lucky Ship' by its crew, who did so well during WWII.*

OPERATIONAL HISTORY

The first section of the keel was laid down on Wednesday 4 August 1971 at N.V. Koninklijke Mij. De Schelde in Vlissingen (Flushing). On Saturday 2 June 1973 she was baptized by H.R.H. Princess Beatrix.

B.V. KONINKLIJKE MAATSCHAPPIJ „DE SCHELDE"
VLISSINGEN / LID RIJN-SCHELDE-VEROLME GROEP

Program for the occasion.

As tradition goes, after the ceremony the axe was sold for one cent. The princess receives the relic from president-director J. Bout.

Tromp *leaving for trials. spotted by a Soviet merchant in 1975.*

Trials commenced on Monday 10 March 1975. Manned by shipyard specialists and crew members she departed at noon. By coincidence a Soviet merchant passed. Nothing special, but there is a Cold War going on and the following day a national newspaper published a photograph with caption reading: Soviet interest for new GW frigate. Confronted with a minor defect on its stabilization system the ship made a short port call on Rosyth. *Tromp* returned to De Schelde shipyard on 20 March.

Acceptance-trials planned from 1 April to 20 May in European waters. First 10 days of April in Den Helder and thereafter proceeding to the Portland exercise areas also negotiating the "noise range". While underway the 'Tromp Zjoernaal' emerged, the on-board news broadcast, that remained active over the years. Test and trials program continued off Norway with visits to Kristiansand, Haakonsvern and Stavanger. 13 - 20 May in Den Helder.

In British waters.

After one week in Den Helder *Tromp* departed 20 May to continue trials in areas with a warmer/tropical climate. Special attention was paid to electronics/weapon systems. On 26 May she visited Santa Cruz de Tenerife. The newspapers published articles about the new frigate named after the famous tactician who defeated the Armada in 1639. A few days later she entered the port of Dakar thereafter proceeding south to test under tropical conditions and returning to Dakar for refuelling. When heading back to the north she encountered stormy weather. At full speed she had to deal with roughers, rolling, pitching and sometimes slamming in a head sea. In Lisbon a swarm lady-bugs took shelter in the ship. After passing the Channel the twin guns were tested proceeding at high speed. Besides a tremendous noise some damage was inflicted at the superstructure and 3D-radar. 16 June arrived Vlissingen; finished the tropical part of the acceptance trials.

In the Canarian newspapers attention for the new ship visiting.

IN SERVICE

On the premises of De Schelde yard *Tromp* was commissioned 3 October 1975 and sailed to Den Helder on 15 October. Leaving on Monday 27 October for exercises and trials. Once at sea the commanding officer of Sqn 860 conducted the first landing with a Wasp helicopter. Following the tradition, he was received with a cake. On Friday 7 November HRH Prince Claus arrived by Wasp helicopter for a visit. Leaving the ship after arrival in home port.

On 24 November to Wilhelmshaven for degaussing. Returning to the yard on 19 December for minor modifications. The year 1976 started with work-up at FOST. Once completed (6 Feb.) the frigate was considered fully operational and ready for duty. Leaving Portland heading south to the Bay of Biscay for rendezvous with the NLTG for a 6 week assignment. During the exercises the gun crew succeeded to impress the Task Group by destroying the target drone with a single shot! ENDEX!

Training

FOST (Flag Officer Sea Training) in Portland (UK) was to certify the crew as being sufficiently prepared for any eventuality through rigorous exercises and readiness inspections. It combined surveys of the physical condition of the ship with tests of the crew's readiness for deployment, including a weekly war-fighting and damage control scenario known as the 'Thursday War'.

In the weekend the ships visited Gibraltar where the CTG and staff embarked. On 5 March the ships arrived in Den Helder for leave. After the weekend the NLTG departed for 3-weeks exercises and visits to Plymouth and Brest.

On 20 April *Tromp* left for the United States to attend the Bicentennial celebrations. Also the loading and testing of guided weapon systems was scheduled. After passing Lands End the weather deteriorated. Squally rain showers soon urged to cancel sports on the helideck. Following day the Tyne cruising turbines failed and the Olympus had to take over. While technicians investigated, the weather went worse and the ship had to proceed with reduced speed. When altering course for the last trajectory, the frigate left the depression. On 29 April 1976 the coastline was sighted and the weather was finally improving. Passing the bridge of Chesapeake Bay a salute was fired to the Admiral Mid Atlantic Region at Naval Station Norfolk. *Tromp* moored, opposite of host ship USS *Josephus Daniels* (CG-27). The *Belknap* class cruiser made quite an impression by offering an extended entertainment program. Few days later she sailed up-river to load missiles. Carrying out exercises and visiting the ports of Charleston and San Juan. Early in the morning of 24 May at the Atlantic Fleet Weapons Range *Tromp* launched the first missile. Tests continued two more days. A total of 5 Tartars and 5 Seasparrows had been launched. Strict safety regulations ordered all hands below deck. They could witness launchings later on video.

Entered Delaware River on 7 June for a visit to Philadelphia. A Netherlands warship visiting was rare and generated much public attention.
After a week she joined the NLTG (*Holland*, *Zeeland*, *Van Nes*, *Van Galen* and *Poolster*) for exercises in the Virginia Capes Operating Area. Later joined HMS *Ark Royal* and USS *Paul*, USS *Jonas Ingram* and submarine USS *Hammerhead*. Then to Norfolk to prepare for the Bicentennial event in New York. On Saturday 3 July *Tromp* sails on the Hudson River as sixth in line, of the 55 warships. While saluting guns roared, she arrived

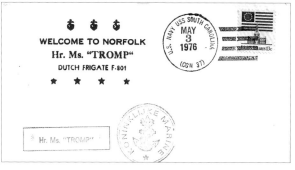

The United States Bicentennial was a series of celebrations and observances that paid tribute to historical events leading up to the creation of the United States of America as an independent republic. It was a central event in the memory of the American Revolution. The Bicentennial culminated on Sunday, July 4, 1976, with the 200th anniversary of the adoption of the Declaration of Independence.

at noon and anchored near the George Washington Bridge. The following day the city really came to live, besides the many celebrations there was a flag parade and the arrival of the tall ships. In the evening warships were illuminated. Four days later *Tromp* departed home bound. After some days a mysterious noise on a prop shaft was recorded. Reason to stop for inspection. Divers came upon a missing protection shield and had quite a job to prevent the situation go from bad to worse. On 15 July 1976 at Den Helder.

The NLTG (*Tromp* flag) departed Den Helder 30 August. After a port call the ships left Rosyth 6 September for a pre-exercise work-up in areas NW of Cape Wrath. 10 September anchoring at Scapa Flow for briefing on board USS *Mount Whitney*. NATO exercise Teamwork '76 lasted from 10 - 23 September.

NL units in
TEAMWORK '76

2 DD - 2 FFG - 5 FF - 2 SS - 1 AO - 7 MPA's - 10 MCM vessels, 2 Survey vessels, 1 diving tender and an amphibious combat group.

Returning in Den Helder.

After port visits to Amsterdam and Rotterdam *Tromp* arrived 30 September at the builder's yard for maintenance.

Most of 1977 *Tromp* acted as flagship for the NLTG. Leaving on 24 January with *Van Nes*, *Groningen*, *Rotterdam* and *Zuiderkruis* for NATO exercise Locked Gate. An exercise aimed to deny the entrance to the Mediterranean in the event of war. To control the Strait of Gibraltar preventing Soviet units to break out into the Atlantic. While proceeding a practice program was carried out. Near the Portuguese coast submarine *Tijgerhaai* joined. (Also submarine *Tonijn* participated.) After the NATO exercise the NLTG visited Toulon (11 to 21 Feb.). Later exercising off Sardinia with units of the Italian navy and US Sixth Fleet. On 22 February a Lynx helicopter landed for the first time on *Tromp*. Visiting Palermo (25 - 28 Feb.) and Naples (4 - 7 Mar.). During this visit 400 sailors went on audition to the Pope Paulus VI. After more exercises the NLTG visited The Rock (Gibraltar (11 - 14 Mar.)). Arrival in Den Helder on 18 March.

Tromp sailed 18 and 19 April showing demonstrations for invited authorities. Assigned to the NLTG and visiting Scapa Flow (23 - 24 Apr.). Joined national exercise Raving Nut before returning to Den Helder (6 May). On 8 June the NLTG paid an informal visit of 5 days to Leningrad (St. Petersburg) and then sailed to Rotterdam. *Tromp* attended the Silver Jubilee Fleet Review (25 - 29 June.) at Spithead Roads.

12 September. Departed for "Autumn" exercises in North Sea/Channel areas. Visits to Torquay and Rouen. *Tromp* entered Portsmouth 29 Oct. for PXD (post exercise discussion) of Ocean Safari (17-29 Oct.) executed to test the ability to supply Europe in a future war.
In Nov/Dec. maintenance and leave.

Tromp was assigned to STANAVFORLANT on 9 December. With staff embarked she left Den Helder on 4 January 1978 and during the dog watch FGS *Emden* reported for duty. Two days later HMS *Naiad* joined and after fuelling at Ponta Delgada the

ships crossed the Atlantic to the Bermuda's arriving 15 January at Ireland Island where STANAVFORLANT assembled. The next day HMCS *Skeena* and USS *Pharris* completed the squadron. En route Charleston, arrival 19 January. Working up in the Jacksonville exercise areas and a visit to Mayport followed. In February at Roosevelt Roads. Before Readex 1-78 (11 - 20 Feb.) started a Tartar and a Sea Sparrow missile were launched to intercept a drone. After Readex *Tromp* berthed in Fort Lauderdale where an "Open House" attracted 5000 visitors in two days to

June 1977. Anchored at Spithead for Silver Jubilee Fleet Review. CO Capt. J.H. Scheuer; flagship of Radm. J.H.B. Hulshof (COMNLTG). Dressed overall with White Ensign at the masthead. Parked on deck the predecessor of the Lynx the one engined Wasp. The first shipborne helicopter of the RNLN, operational between 1966-1981. All helicopters were assigned to Squadron 860.

admire the Dutch Guided Missile Frigate. Successful "Holland Promotion"!
After exercise Safe Pass the ships entered Norfolk on 16 March. One week later STANAVFORLANT arrived in Halifax (N.S.) and on 31 March command was delegated to the Canadians (HMCS *Iroquois*). The following day *Tromp* sailed to Mayport, negotiating heavy weather, to join the NLTG.
On 12 May 1978 the TG arrived in the home port. *Tromp* sailed to Vlissingen 14 August and entered De Scheldepoort dockyard the same day for a periodic (2-yearly) regular maintenance .

After work-up early 1979, the ship was assigned to the NLTG as flagship for a journey to the Middle- and Far East. Leaving on 12 March along with *Drenthe*, *Kortenaer* and *Poolster*. Making port visits to Tanger and Alexandria. When *Poolster* and *Tromp* entered the Suez Canal they were the first ships of the Neth. Navy passing since the closing of 1967. Visiting Jiddah (3 - 5 Apr.) *Drenthe*, *Kortenaer* visited Alexandria. Leaving the Red Sea the ships made a rendez-vous with French units in the Gulf of Aden. Busy Passex was carried out with FS *Duquesne* (D603) -also with large radome-, FS *Bouvet* (D624) and *La Charente* (A 626) a tanker and commandship of the French Indian Ocean Forces. Sunday 8 April at twelve hundred hours the Passex was completed 13 N - 50 E and the French returned to Djibouti. In Bombay (12 - 16 Apr.) there was an exhibition of Dutch industrial achievements on *Tromp* and *Kortenaer*.

On 22 April *Kortenaer* and *Drenthe* entered Port Kelang the harbour of Kuala Lumpur while *Tromp* and *Poolster* sailed

to Singapore where the Task Group commander paid his respects by laying a wreath at the memorial and Netherlands war graves. Some days later the NLTG reassembled at sea and proceeding to Indonesia. Underway again the ships crossed the Line and King Neptune and his suite of "shellbacks" boarded. Many a "polliwog" shivered by the sound of the boatswains call; the majority of the crew had to be baptized! (26 Apr.) Executed Exercise Holindo II with Indonesian DE's *Samadikun* (341) and *Martadinata* (342) in the Java Sea (27 Apr.).

28 April, arrival in Tanjung Priok the harbour of Jakarta. In the former colony respects were paid to the fields of honour

1979; Far East Australia cruise, sailing to the Middle and Far East.

of Kalibata and Menteng Pulo. On the quay there was on 30 April a ceremony performed to celebrate the 70th birthday of HM The Queen. After embarking the ambassador and ten Indonesian naval officers the Task Group left for visiting Surabaya (3 - 8 May). Respects were paid by laying wreaths at the cemetaries of Kusuma Bangsa (Ind.) and Kembang Kuning (Neth.) After leaving Surabaya, the CTG held a memorial in Java Sea, for the 2300 servicemen wgo lost their lives. Slowly proceeding with 100 yards distance, the crews manning the rail and the CTG committed a wreath from *Kortenaer* to the sea.

Cold War at sea 1945-91

Tromp had been built during the Cold War. Frigates were employed in Blue Water Ops to protect the Atlantic Lifeline. Annually millions of tons of food and raw materials cross the Atlantic. Danger areas like the Western Approaches to the British Isles demanded defence.

The GIUK* gap needed ASW units to deny the Soviet submarines entering the Atlantic. After 1991 the threat seemed gone, but 30 years later the navy is nearly knocked out due to the so called "peace dividend" policy.

E.g. in 2006 all excellent and updated Orion MPA's were flogged. The Neth. navy, once the largest of the "small NATO navies", now neglected and below NATO standard.

_____
* Greenland-Iceland-United Kingdom

After passing Bali Strait the seastate increased. Australian ships came out to welcome the Dutch and to execute exercise Dirk Hartog (11 - 18 May). Aussie ships *Perth* (D 38), *Vendetta* (D 08), *Derwent* (D 49), *Acute* (P 81), *Orion* (S 61) and also Orion MPA's and F-111 fighters took part. Entering Fremantle on 18 May to celebrate WAY 79 the 150th anniversary of the European colonization of Western Australia. Excitement occurred when some naval veterans reported on the gangway. They were former crew of cruiser *Tromp* who served during WWII and now were living in Australia. They presented the war flag of the cruiser as a gift for the navy museum. Ceremonies at the Submarine and State War memorials; Karrakatta cemetery (25 Neth. naval airmen Broome Incident victims) and a plaque near Cape Leeuwin. From 1-5 June

Drenthe visited Geraldton. A wreath was laid at the Batavia Memorial. With invited guests embarked a trip was made to Beacon Island. There the NL ambassador unveiled a plaque which was placed to commemorate the loss (4 June 1629) and the infamous violent mutiny by crew of the Dutch East India man *Batavia*. On 5 June the TG sailed via Diego Garcia (14 - 16 Jun). Again a Passex with French units and later in the Red Sea with USS *Sampson* (DDG 10) while *Poolster* loaded oil in Jiddah. Other ports of call: Istanbul (*Tromp*, *Drenthe*) and Piraeus (*Kortenaer*, *Poolster*). On Thursday 12 July the cruise had been completed.

On 10 September the TG departed (*Tromp*, *Utrecht*, *Kortenaer*, *Zuiderkruis*) being reinforced by the new Belgian *Wielingen* (F 910) and *Westdiep* (F 911). After a visit to Rosyth (14 - 17 Sep.) in the major exercise Ocean

Safari (24 Sep. -5 Oct.). The feasibility of shipping supplies and reinforcements from the US to Europe was being tested in the exercise by "attacks" from submarines, planes and surface ships. Involving more than 17,000 men, 70 ships and 200 aircraft. Returned in Den Helder on 6 October. On 22 October the Midshipmen cruise started. The TG (minus *Westdiep*, plus *Zeehond* (S 809)) sailed south to find friendly weather exercise areas. Ports visited: Lisbon, Funchal Gibraltar. 22 November arrival in Den Helder.

EIGHTIES

22 to 24 January *Tromp* sailed for individual SEWACO trials (sensors and command systems). A few days later (28 Jan.) the NLTG sailed for the Mediterranean to participate in the ASW exercise Dogfish. Also *Callenburgh*, *Zwaardvis*, Dutch MPA's from Sigonella, Augusta, Sicily and USNS *Mississinewa* (T-AO 144) joined the TG together with units of the US 6th Fleet. Port visits were made to Palermo, Toulon, Barcelona and Valencia. Launched missiles on the range of Levant Island. Returning on 6 March.

The TG departed 14 April. On 15 April many ministers of the NL cabinet embarked in *Tromp* and *Callenburgh* to witness an impressive show of readiness. WPP (Weekly Practice Program) crammed with exercises e.g. ASW with *Zeehond*; the Schreiner Company launched drones. Ports of call Lisbon, Tanger. 14 May Spring Voyage (1) completed.

Executing RAS with HNLMS Zuiderkruis

Order of the Blue Nose

A Navy tradition which dictates that when Sailors cross into the Arctic Circle, above 66°34'N. they enter the realm of Boreas Rex, King of the North. The only way to be accepted into the order is to successfully complete his list of challenges.
Following the tradition, the youngest sailor and officer, both dressed in tropical kit were called to paint the 'Blue Nose' (Panama hawse painted blue).

The badge 'Netherlands - USA 200' was created by Gert Dumbar. An American 5 pointed star merging in a flag with 7 bars. Symbol of the 7 provinces of 1782. The wave stands for the ocean between both countries. The circle is symbol for friendship. At the base 13 stars, for the states of 1782.

Jun 2 - 17. Second Spring Cruise; special attention to Air Defence. Visits to Guzz and Pompey (RN slang for Plymouth and Portsmouth). Participated in Navy Days 1980 before receiving scheduled 2 years maintenance. On 23 June, *Tromp* was the first to enter the newly constructed roofed Frigate Dock (Dok 6) of the 'Rijkswerf' (naval dockyard) for maintenance. It was the opportunity to replace the large dome. It sustained damage when a F-14 in low level flight passed too close with high speed during an exercise.

From 12 - 15 December *Tromp* in company with *Poolster* and *Groningen* berthed at Rotterdam to celebrate 315th anniversary of the Marine Corps (10 Dec. 1665).

On 26 January 1981, the NLTG (*Tromp* (flag), *Kortenaer*, *Van Kinsbergen*, *Poolster* and submarine *Dolfijn*) departed for the Mediterranean. In Channel areas practised CASEX (ASW) and ADEX (AAW) to prepare skills for the next NATO exercise. A visit to Plymouth (crews always enjoyed British ports). Then to the Gibraltar areas to partake in the Test Gate exercise (5 - 12 Feb). Making port calls to Cadiz, Casablanca, Las Palmas and Funchal. On 13 March reaching Den Helder again.

Spring cruise '81 (21 Apr. -27 May)

TG 429.5: *Tromp*, *Zuiderkruis*, *Van Galen*, *Van Speijk*, *Banckert*. ASW exercises in Skaggerak. A Danish submarine played the "Clockwork Mouse". Joined by *Van Nes* and *Piet Heyn* (from Frigate Squadron) happy to hone their ASW ability. Launching Seasparrows against target drones provided by the Schreiner Company. On 14 May the Polar Circle was crossed. Ports of call: Copenhagen, Bergen and Tromsö.

Departed 9 June the TG with *Tromp*, *Zuiderkruis*, 2 Leanders and 3 S-class frigates executed a special practice program"Gold Nut" as a prelude to NATO exercise Roebuck 1981 (12 - 19 June).

Summer Cruise (26 Aug - 17 Sep.)

The Dutch attended the British Navy Days. Portsmouth *Tromp*, *Zuiderkruis*, *Van Speijk* (47.622 visitors) and Plymouth *Callenburgh*, *Van Nes* (15.000).

In NATO exercise Ocean Safari (8 - 18 Sep.) the role of the TG was restricted to play "Orange Forces". 19 Oct - 13 Nov.: Autumn Cruise, Junior Midshipmen embarked. Units: *Tromp*, *Zuiderkruis*, *Van Speijk*, *Van Nes*, *Callenburgh*. Sparring partners, MPA's: US Orions, UK Nimrods, submarines *Tijgerhaai* and *Zeehond*. Ports of call: Lisbon, El Ferrol, La Coruña and Plymouth.

Assigned to the TG in 1982: *Tromp* (flag), *Van Speijk*, *Callenburgh*, *Piet Heyn*, *Wandelaar* (BE), *Overijssel* and *Zuiderkruis*. Departing for the United States (8 Feb.) to celebrate the 200 years diplomatic relation between the USA and Netherlands. While crossing the Atlantic the weather rapidly deteriorated. Near the Azores a distress call was intercepted (12 Feb.). Transmitted by the Greek tanker *Victoria*, with a cargo of 22.000 ton molasses on her way from Florida to Liverpool. *Van Speijk* and *Callenburgh* both carrying helicopters were dispatched, despite the distance of 400 nm in poor weather to the rescue. Arriving at dawn to find the still floating after part of the merchant pitching to 7 meters (23 ft)

During Navy Days. Note the colourful badges on the covers of Seasparrow launcher. The Netherlands-USA 200 badge of previous page can be spotted on the SB canister.

and rolling up to 45°. High wind 30 kts - Beaufort 7- Both helicopters rescued a total of 16 crew; 16 others did not survive.

Later *Tromp* encountered problems with the SB stabilization. An inspection revealed a missing protective cover. Therefore, she docked in Willemstad Curaçao (25 Feb.). The repairs took only 3 days.

Several exercises where executed in areas near Roosevelt Roads where submarine *Zeehond* joined to act as opponent. After-

wards the ships headed for Florida to visit Fort Lauderdale (5 - 8 Mar.) and Norfolk (19 - 29 Mar. and 11 - 12 Apr.).

On 13 April *Callenburgh* and *Tromp* paid a visit to Baltimore. The NLTG gathered again on 20 April to proceed to New York to add lustre to the visit of HM Queen Beatrix and Prince Claus to the USA. After the warm welcome in New York Harbor by spouting tugs, the ships moored on 22 April. Some days later the royal couple visited

Tromp preparing to moore alongside of De Ruyter, *in the background supply ship* Poolster.

the Task Group. Leaving on 27 April for exercises with Canadian units and a visit to Halifax (6 - 10 May) before returning to Den Helder.
On 19 May the TG arrived after a 100 days journey, the log recorded 18.332 nm. *Tromp* was relieved as flag ship on 1 July by sister *De Ruyter*.

After summer leave and taking part in exercises Silver Nut (30 Aug. - 3 Sep.) and along with *Wandelaar* (F 912 - BE) in Northern Wedding (6 - 17 Sep.).
A maintenance period followed, to update the SEWACO command system and receiving the first Link 11 communication system.

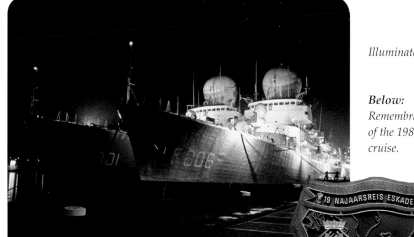

Illuminated sisters.

Below:
Remembrance crest of the 1984 autumn cruise.

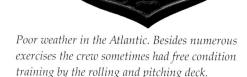

Poor weather in the Atlantic. Besides numerous exercises the crew sometimes had free condition training by the rolling and pitching deck.

Working up with a fresh crew (3 - 24 Feb. 1984). The state of readiness of *Tromp* was satisfactory and on 9 April the flag of COMNLTG was hoisted.

Spring Cruise (24 Apr. - 25 May)
Tromp, Zuiderkruis, Banckert, Pieter Florisz, Evertsen, Van Nes later joined *Isaac Sweers* with midshipmen. Exercises west of Portugal. NATO exercise Open Gate 84 (27 Apr. - 3 May) Ports of call: Gibraltar and Las Palmas.

Autumn Cruise (22 Aug. - 17 Nov.)
Tromp, Poolster, Piet Heyn, Pieter Florisz, Bloys van Treslong, Evertsen, Isaac Sweers, Potvis. NATO exercise Safe Pass 84 (10 - 21 Sep.). Executing a busy Weekly Program the TG sailed via Pentland Firth to arrive in Reykjavik on 30 Aug. Then to Nova Scotia to visit Halifax (7-10 Sep.) "Safe Pass"was interfered by hurricane Diana a tropical cyclone which hit the US East Coast. The NATO ships steered evasive

courses and -in particular- air defence exercises were cancelled. The eye of the cyclone passed at 140 nm distance. Still the ships experienced force 8 / 9 at least. In Safe Pass offensive exercises were carried out by STANAVFORLANT, Canadian and USN units. *Tromp* stayed in Norfolk (22 Sep. - 8 Oct.) for maintenance and testing the Link system. From 8 October the CINCRNLN* embarked *Tromp* for inspections until 12 October.
TG gathered again at sea for exercises off the coast of Virginia. Divided in groups, the ships visited the Leeward Islands, Curaçao, Aruba and Bonaire (19 - 22 Oct.) and La Guaira, the port of Caracas. Subsequently, *Tromp* carried out a gunnery practise on the island of Vieques and launched a Standard missile.
The Windward Islands were visited (2 - 5 Nov.) followed by exercising in the areas off Puerto Rico. The TG returned via the Azores to Den Helder, arriving on 17 November.

Approaching RFA Olna for RAS

---
* = Vice Admiral J.H.B. Hulshof

HRH Prince Willem Alexander served several months aboard Tromp.

Right:

The commissioning ceremony on the heli deck. ADMIRAL-NLFLEET Vice Admiral R. Krijger and the prince watched by his parents. Painting by Rob Hessels.

Early 1985 the TG consisted of *Tromp* (flag), *Evertsen, Isaac Sweers, Callenburgh, Van Kinsbergen, Piet Heyn, Zuiderkruis, Potvis, Tijgerhaai,* 4 Lynx helicopters and NL Orion MPA's operating from Sigonella. For the Winter Cruise (7 Feb. - 20 Apr.) with a combined crew of about 1600 officers and ratings to the Mediterranean for ASW exercise Dogfish. The ships visited (Showing the Flag): Venice, Catania, Heraklion, Alexandria, Haifa, Tunis, and Alicante.

The AAW exercise Iron Nut was carried out north of the Frisian Isles; ready duty ship *Tjerk Hiddes* reported happy to participate. The TG proceeded to Rosyth invited to show the flag during the Navy Days (7-11 June). Then in North Sea and West of Scotland exercise JMC/Roebuck until 24 June. *Poolster* mothership of two RN Sea King helicopters.

Autumn Cruise (26 Aug. - 25 Oct.)
(*Tromp* (flag), *Van Nes, Kortenaer, Banckert,*

Safari southwest of England with *Potvis* and *Zeehond.* On 18 September, the crew witnessed a firepower demonstration of battleship USS *Iowa* (BB 61).
Leixoes was visited and subsequently submarine *Potvis* emerged as sparring partner in the Bay of Biscay. The NLTG was split up for visits to Bordeaux and La Rochelle and *Tromp* carried out tests with the embarked Lynx helicopter. Exercises in the North Sea with NLFREGRON and STANAVFORLANT. Individual ship visits to Leith/ Dundee/ Rosyth.

Departed 11 January 1986 bound for Charleston to join STANAVFORLANT. Among the crew was HRH Prince Willem Alexander who was commissioned on 1 July 1986 as Sub-Lieutenant KMR (Royal Naval Reserve)[**].
13 January RV with RFA *Green Rover* (A 268) and 22 January a RAS with USNS *Kalamazoo* (AOR 6) and reaching Charleston next day; until 28 January. Sailed to Mobile in company with HMS *Battleaxe* (F 89) ASW exercises with FGS *Hessen* (D 184), USS *L. Mendel Rivers* (SSN 686) and HMCS *Ottawa* (229).
While sailing *Tromp* suffered of a malfunctioning compass. Steering orders were passed and executed by way of the SCC (Ship's Control Centre). The technical branch was able to fix the problem.

Visiting Halifax in 1987. HNLMS Tromp, *a Canadian frigate and HMS* Danae,

Spring Cruise (28 May - 26 Jun.)
TG units: *Tromp, Poolster, Kortenaer, Van Kinsbergen, Evertsen, Philips van Almonde, Wandelaar* (BE) and 4 Lynx helo's. A visit to Rotterdam exempt *Tromp* acting as host ship during the change of command by the CINCRNLN in Den Helder (1 June).

Poolster (again with two Sea Kings). and *Isaac Sweers, Evertsen, Callenburgh* until 20 September and for some weeks *Wielingen* (BE), *Wandelaar* (BE). After work up the ships participated in exercise Botany Bay escorting a Denmark bound convoy. Followed by exercise Ocean

---

** First entry recorded in the Navy List: Temp. Sub Lt SD (Special Duty) KMR (RNR) = I April 1986

In Mobile the ship welcomed over 1600 visitors during 'Open House' (9 Feb.). Underway to St. Thomas on 13 February a small fire broke out in the aft engine room. Rigid damage control drills paid off: the fire was quickly extinguished! After the visit to St Thomas (17 Feb.), the ships went to Roosevelt Roads, Norfolk, New York,

St. John, Halifax, Ponta Delgada, Brest, Bergen, Antwerp and Lisbon and Den Helder (16 June).

In company with *Piet Heyn* and submarine *Tijgerhaai*, leaving port on 3 July for Kiel and exercises in the area. Once returned *Tromp* received regular maintenance.

Above:
Visiting Baltimore.
HMS Danae *is*
moored alongside
Tromp.

Route NLTG 1989.

Launching Tartar.

Departing on 27 October, for work up, visiting Rosyth and Joint Maritime Course exercises. Visiting New Castle (14 Nov.) before returning.

On 8 January 1987 *Tromp* was assigned flagship for STANAVFORLANT. En route to the Azores with FGS *Rheinland-Pfalz* (F 209) and HMS *Danae* (F 47). Midway the crossing to Halifax (NS) USS *Taylor* (FFG 50) and HMCS *Assiniboine* (234) joined. Encountering stormy weather two times and some damage was inflicted. The ships arrived in Halifax on 22 January. A few days later the ships went south to Virgin Islands, visiting St. Thomas (6 - 9 Feb.) and joined the major American exercise Fleetex 1-87. Port visit to Fort Lauderdale (25 Feb. - 2 Mar.). Underway to Baltimore (to stay 6 - 12 Mar.) ASW exercises with USS *Silversides* (SSN 679) were executed.
Fund raising for charity! The cycling team of *Tromp* achieved a 7 days lasting Tour from Baltimore to Norfolk. This Charity Cycle Ride collected $ 4000.- Berthed at Norfolk (13-28 Mar.) "The biggest Naval Installation in the World"... the USN mates claimed! On 27 March *Tromp* was relieved in SNFL by *Van Kinsbergen* (F 809) and departed next day to join the NLTG on 1 April. Two months exercises followed. Three port vistis were made; to Oranjestad (Aruba) with *Pieter Florisz*, to Wilmington (NC) with *Jan van Brakel* and to Philadelphia with *Banckert*.

On 28 May the ships arrived in the homeport. On 14 August after summer leave the scheduled multi-yearly maintenance commenced.

On 6 June 1988 *Tromp* started a period of trials and training. As usual this demanded a dedicated crew to achieve operational readiness. Most eye catching addition were the SCOT*** SHF domes each side of the 3-D radar.

In July 1988 *Tromp* and *Harlingen* (M 854) escorted about 200 yachts from Hellevoetsluis to Torbay to celebrate "The Glorious Revolution"of 1688. The Dutch "Stadtholder" Willem III was married to Mary Stuart II. In 1689 he was crowned King William III ("King Billy"). During the tour/race to Brixham (Torbay) the escorting ships rendered assistance to many participating vessels of which some took shelter in various ports after struggling against a SW-gale (7). Also in Torbay: HMY *Britannia*, HMS *Exeter* (D 89) and *Amazon* (F 169). A Grand Sail Past happened, about 300 vessels brought their salute to Queen Elisabeth II, Prince Philip and Prince Willem-Alexander.

Tromp participated in Mini Navy Days (23 - 25 Sep.) in Rotterdam. Over 15.000 people visited the ship tasting life in the navy. In 1988 *Tromp* paid visits to: Kiel, Aarhus, Brest, Cowes, Weymouth, Torbay (Jul.) Portland (Okt. - Nov.) and Antwerpen (Dec.).

*** SCOT = Satellite Communications Onboard Terminal

Attending Navy Days in Den Helder.

On 27 February 1989, the TG consisted of: *Tromp* (flag), *Jacob van Heemskerck*, *Kortenaer*, *Callenburgh*, *Witte de With*, *Poolster* and *Zuiderkruis*. The ships departed for work up and taking part in the major NATO exercise North Star (5 - 15 Mar.) in the northern part of the North Sea and Norwegian Sea. The US Carrier Battle Group with USS *America* (CVA 66) was also present. It was followed by a port visit to Rosyth, missile launching on the range near Bordeaux and a visit to Lisbon (31 Mar.). Took part in the Spanish ASW exercise Tapon '89 (4 to 12 Apr.) in Gibraltar Strait. Visiting Cadiz (13 - 17 Apr.), Portsmouth (4 May) and AAW exercise Square Nut; in 3 days time about 50 various aircraft **** 'attacked' the TG. *Poolster* operated communication jammers, trying to deny message traffic.

**** F-16, Phantom, Mirage, Buccaneer, Tornado, NF-5 and Sea Harrier.

In July *Tromp* participated in Navy Days.

TG cruise 2/89 (28 Aug. - 22 Sep.)
Tromp, Zuiderkruis, Kortenaer, Banckert, Philips van Almonde, Callenburgh, Abraham Crijnssen, Van Kinsbergen with 4 Lynx helo's. On 28 August the TG had a 'silent departure', leaving port with radio silence imposed. All signals by flashing light. In the North Sea a short work up followed, passing Dover Strait to the exercise areas. Joining the major NATO exercise Sharp Spear 89 (8 - 21 Sep.) off the south coast of England and Plymouth involving 270 ships, 300 aircraft and 40.000 officers and ratings. The setting of this exercise was defending the shipping lanes of North Sea ports in shallow waters. After a few days the exercise scenario continued with a "free phase" allowing CO's more freedom to act independent in response to various threats.

Executing RAS

1990: Tromp, Poolster *and* Jacob van Heemskerck *in the Norwegian Sea.*

Arriving in the "States". Tromp just berthed, the large harbor tug, 2000 bhp, Oshkosh (YTB-757) assist a S-Frigate.

The NLTG had rendezvous with the British TG north west of Ireland and escorted a convoy from Loch Ewe to the Norwegian coast. After the exercise *Tromp* proceeded to IJmuiden to embark several political VIP's for a cruise to Rotterdam.

When on 22 October the TG departed the helicopters were absent due to problems with controls, all Lynx helicopters were grounded. Nevertheless, the ships went for ASW and AAW exercises south west of Norway. Visiting Aalborg, Copenhagen and Stavanger followed by exercises at Devil's Hole, an area off the Scottish east coast and visits to Leith and Rosyth. Between 14 and 24 November *Tromp* executed the Joint Maritime Course. Now with ships helicopter since the Naval Air Service had resolved the problems. The TG returned to Den Helder on 28 November.

NINETIES

TG Cruise 1/90 (22 Jan. - 18 May)
In 1990 TG was assigned for a journey to South America called 'Fairwind '90'. Initially with *Tromp* (flag), *Jacob van Heemskerck*, *Bloys van Treslong* and *Poolster* with Lynx No 66 *Tromp*, Lynx no. 270 *Bloys*. Frigates *Pieter Florisz* and BNS *Wielingen* joined on 27 April.

Entering the British Channel a storm was encountered but exercises continued and the French aircraft proved to be capable opponents. By the time the Portuguese exercise area was reached, the exercise with Portuguese aircraft was cancelled. The ships sailed to Lisbon were submarine *Tijgerhaai* joined.

Looking forward on J-deck. The reinforced platform above is supporting Harpoon.

Left: Eyeball Mk.1 port side optical sight. The binocular was stored and could easily be mounted.

In São Vicente, Cape Verde *Tromp* berthed in Porto Grande Mindelo with 2 frigates alongside and *Poolster* berthed with *Mercuur* and *Tijgerhaai* alongside. Three days later departing and exercises with *Mercuur* (A 900) and the submarine. Crossing the Line, on 24 Feburary, entering King Neptune's domains. He arrived with members of his Court and all novices (polywogs) were lathered, shaven and ducked. As a persiflage on Artic Circle "Blue Nose" the Panama hawse hole was painted red. Arriving a few days later in Rio de Janeiro (1 March). One of the days in port was 'open house', were many Brazilians took their opportunity. Duty personnel at the gangway were surprised not to see only handbags given in custody but also some guns!

The ships departed on 6 March and while executing the Weekly Program a high speed of advance was maintained, course to the Leeward Antilles. Crossing the equator the second time King Neptune appeared again because his policemen had caught a lying sailor who claimed to be a "shellback". Neptune was annoyed and ordered a special treatment...
Underway about 25 sailors applied for a crosspol. They were transferred by light jackstay. On 15 March the TG passed the Galleons Passage, gate to the Caribbean.

Two days later the Defence Minister landed by helo on *Tromp*. "Heemskerck" and "Bloys" visited Oranje-stad, Aruba from 16 - 19 March the other ships visited Willemstad. To their surprise family and relatives were waiting when the ships berthed. Later the 2 FF's joined the others in Willemstad. A week of maintenance and liberty followed while the technical branch replaced the cruising turbines. Along with ordered parts some yard technicians arrived by plane. On 26 March *Tromp* was ready. The WIG***** *Banckert* (F 810) joined 28 March and after exercises with the Venezuelan navy La Guaira was visited.

Crosspol

The opportunity for crewmembers to exchange workplaces and serve for a few days on another ship. The other surroundings and sometimes different equipment is always considered as a valuable experience.

On 2 April the ships sailed to an area off Puerto Rico and exercised with planes of USS *Saratoga* (CV-60) and visited Roosevelt Roads to load missiles for launching at the missile range before returning to the Netherlands Antilles.

***** WIG = West Indies Guardship

With some delay (engines *Poolster*) the TG left the Bay of St. Maarten on 18 April 0800 lt -homeward bound. Added to the WP was a PASSEX with German ships underway to the Caribbean. On 25 April the "Bloys" missed a man. A thorough search alas had no result.

In Funchal (27-30 Apr.) the frigates *Pieter Florisz* (with Lynx 278) and BNS *Wielingen* joined. After the weekend the ships sailed to the exercise area west of Gibraltar Strait for NATO exercise Open Gate (1 - 11 May) with about 6000 men in 36 warships, 10 merchants and 57 aircraft involved. "Man lost overboard" a distress call from Spanish fisherman *Aquilica*. The helo search had a negative result. Last port visit of Fairwind '90 was Portsmouth (11 - 14 May) before participating in exercise Hard Nut north of the Frisian Islands. A farewell to COMNLTG was given by a Sail Past. After a very satisfactory 17 weeks exercise cruise the TG arrived in Den Helder on 18 May 1990.

Departing for Fairwind 1990.

In August 1990 started the Gulf War with Operation Desert Shield. To support the Coalition *Pieter Florisz* and *Witte de With* arrived in September, reporting for duty in the Gulf area. This resulted in a smaller NLTG for the Cruise 2/90 (20 Aug. - 20 Sep.) Now with *Tromp* (flag-Lynx 266 embarked), *Piet Heyn*, *Banckert*, *Poolster* (from 27 Aug.- 2 Sea Kings embarked), reinforced with Belgian frigates *Wielingen* and *Wandelaar*. Once at sea (20 Aug.) course was set for the Skagerrak to partici-pate in the annual exercise 'Danex '90' in which STANAVFORLANT and a Danish Task Group also took part. Afterwards *Tromp* and *Banckert* went to Kristiansand while *Piet Heyn* went to Stavanger. The TG was occupied with Teamwork 90 MOD (6 - 17 Sep.) intended to rehearse the wartime reinforcement of Norway.

WAKE UP CALLS...

Oct. 1967 "Missile Age at Sea" started by sinking of INS Eilat off Port Said by Egyptian (Soviet built) missile boats .

Oct.2000 "Asymmetric Warfare": Attack on USS *Cole* (8300 tons -fl-) by a small boat while oiling in Aden harbour.

These attacks boosted decisions for urgent development of new weaponry and tactics!

TG Cruise 3/90 (5 - 23 Nov.)
Leaving port (5 Nov.) now consisting of *Tromp* (flag Lynx 266 embarked), *De Ruyter* and *Banckert*. *Callenburgh* joined a week later. Individual exercises were carried out in the Portland exercise areas. Including noise measurements and high-speed tests. Submarine *Potvis* joined later. The weekend (16 - 19 Nov.) was spent in Plymouth. COMNLTG flew by helo to Portland for a well-wishing *van Heemskerck*, *van Almonde* and *Zuiderkruis* 'en-route' to Persian Gulf. After a few more days of practice, the ships returned to Den Helder.

A 20 mm for short range defence / patrol duties was added in the nineties.

TG Cruise 1/91 (18 Mar. - 12 Apr.)

Tromp (flag), *Poolster, Abraham Crijnssen, Kortenaer* and *Jan van Brakel.*

Only *Tromp* and *Kortenaer* visited Dublin (28 Mar. - 2 Apr.) other ships to Plymouth. Sailing in column the TG presented a ceremonial entry (21 gun salute) to Brest (5 - 8 Apr.). Exercises with FS *Aconit* (F 609), French planes and MCM ships. *Jan van Brakel* streamed towed-array for evaluation with submarine *Tijgerhaai.* 9 April TG exercise Turbulent Tuesday. "Various warfare types" and damage control.

TG Cruise 2/91 (27 May- 4 July)
Midshipmen Cruise

Tromp, Poolster, Abraham Crijnssen, Jan van Brakel, Pieter Florisz and *Potvis.* Midshipmen training ship HMS *Bristol* (D 23) and *Willem van der Zaan* (F 829) arrived for work up. AAW exercises with NATO airforces. Steaming 14 hours through snow covered fjords paying a courtesy visit to Oslo (7 - 10 Jun.). The saluting battery

fired 21 rounds in honour of the Norwegian flag. After exercises in Skagerrak a visit to Aarhus and then to the Baltic (17 Jun.) for annual exercise BALTOPS. On 21 June after a 21 gun salute berthed at Lange Linien, Copenhagen. A visit to Stockholm (28 Jun. - 1 Jul.). Finally, Family days on 4 - 5 July marked the end of this busy cruise.

TG Cruise 3/91 (26 Aug. - 24 Sep.)

Tromp, Poolster, Pieter Florisz, Kortenaer and 3 other frigates only for some days. Commenced training for AAW controllers and OPS-room officers. Visiting Kristiansand (30 Aug. - 2 Sep.); *Poolster* to Stavanger. Exercises Strong Nut (2 - 5 Sep.); 6 September Den Helder. Exercise Northstar (10 - 19 Sep.) The TG was ordered to protect USS *America* (CV-66) against air attacks and submarine threat

on her transit to Norwegian waters. On 15 September, the arctic circle was crossed and Boreas welcomed before reaching Bergen.

TG Cruise 4/91 (28 Oct. - 23 Nov.)

Tromp, Poolster, Piet Heyn, Philips van Almonde, Van Kinsbergen and *Witte de With.* On 4 November departing for Rosyth and Leith before attending the Joint Maritime Course 91/3. STANAVFORLANT also participated. COMNLTG was OTC* over all (± 25) surface ships. Bad weather forced to cancel some exercises. Staff embarked in *Poolster* to Rosyth for PXD** on 25 November and arrived 2 days later in Den Helder.

TG Cruise 1/92 (24 Feb.- 25 Mar.)

Tromp, Zuiderkruis, Witte de With, Banckert, Philips van Almonde.
Exercises in Devil's Hole area with sub *Potvis* and NL Orion MPA's***. Harbours: 28 Feb. - 2 Mar. Newcastle-upon-Tyne and Rosyth (*Poolster* only). Then Teamwork 92 (6 - 21 Mar.) in N-Atlantic and Norwegian Sea.

---

*   Officer in Tactical Command

**   Post Exercise Debrief

*** Maritime Patrol Aircraft

First mission was to intercept a Strike Fleet crossing the Atlantic with USS *Dwight D. Eisenhower* (CVN 69). On the First Watch of 11 March a succesful attack was delivered. Bad weather and ice rind on deck was a menace and *Tromp* had to shelter for a day in the Trondheim fjord.

TG Cruise 2/92 (21 April - 15 July)
Columbus Voyage

Tromp (flag), *Witte de With*, *Van Kinsbergen* and *Zuiderkruis*.
Exercises with sub *Zwaardvis* and two NL Orion MPA's to support a class of OPS

In US territorial waters. Pilot embarked, underway to pay a port visit.

room officer students. In Ponta Delgada 7 - 8 May. In Roosevelt Roads 15 - 18 May, ADMIRALNLHOME embarked *Tromp* to inspect the TG. The *'Stationsschip Netherlands Antilles' Kortenaer* (Guardship) joined 15 - 27 May. Exercise May Fly 92 with USN and German units (17 ships in total). *Tromp* and other ships launched missiles on target drones. Virgin Islands, St. Thomas and St. Croix, were visited and San Juan 29 May- 1 June. Followed by a visit to Willemstad where new guardship *Philips van Almonde* joined. Exercising for two weeks with USN subs and planes. The TG was reinforced by two land-based Netherlands Orion MPA's. The training was postponed for a visit to Fort Lauderdale. There *Philips van Almonde* left the TG. Visiting Baltimore (25 - 29 Jun.) and Boston (1 – 6 Jul.) before sailing homeward bound.

After arrival in Den Helder (15 Jul.). *Tromp* received regular maintenance.
By the end of September *Tromp* was 'chartered' one week to perform PR-trips and entertained about 220 guests a day. From 12 - 16 October certificate qualifications for helicopter direction officers (OPS-room) and flight deck officers (Marshallers). Working up with the frigate squadron (19 Oct.) and a visit to Plymouth (23 - 26 Oct.). From 2 - 16 November taking part in JMC 92/3. 18 November change of command in Den Helder and preparations for STANAVFORMED. When technical issues with 3D radar occurred, plans shifted to repairs.

In March 1993 the frigate sailed again, visiting Malmö (19 - 21 Mar.).

A demonstration day for VIP's caused "operation deep-clean". The Council of Ministers would find *Tromp* spick-and-span as they departed Hoek van Holland on 12 June. The Frigate Squadron conducted the midshipmen cruise from 7 June - 16 July. Units: *Tromp*, *Willem van der Zaan*, *Tjerk Hiddes* (until 18 June and *Karel Doorman* (until 18 June and 5 - 16 July). Also *Poolster* participated. Ports of call: Cadiz, Funchal, Vigo, Dublin, Stavanger.

Ammunitions and Stores were landed. A 'multiple year'-maintenance followed by the Dockyard in Den Helder. It started on 27 September 1993 and would last till November 1994. On 21 November *Tromp* departed, visited Kiel and returned 30 November.

The first months of 1995 *Tromp* stayed some days in Dock no. VI. a.o. for 3D radar repairs. The radar antenna was landed for repairs, after which the ship continued equipment tests. In February working up with FREGRON (Frigate Squadron) visits to Bergen (17 - 20 Feb.) and Göteborg (24 - 27 Feb). Also, the problems with the covers of the stabilization system returned. An inspection revealed some cracks in the support beams. Examination concluded it would not affect her operational status. *De Ruyter* suffered more serious problems, so sister *Tromp* was directed as substitute to visit Russia. representing the Netherlands government. *Tromp* continued to work up (18 Apr.) and carried out exercise Marvika

TG Cruise 2/93 (13 Apr.- 7 May)
Tromp (flag), *De Ruyter* (until 26 Apr.), *Philips van Almonde* and *Zuiderkruis*. Exercise Dragon Hammer in the "Med."cancelled due to situation in former FRY.**** Exercises in The Channel,- *Zuiderkruis* base for two RN Sea Kings-and off the Portuguese coast. Also *Zwaardvis* and Rota based NL Orions. On 22 April the FF's launched Sea Sparrows on drone target planes. Ports of call for *Tromp*: Lisbon (16 - 19 Apr.) and Cadiz (30 Apr.- 3 May). A cinema film titled "Taak voor de Vrede" (A mission for Peace) had been recorded by the Audio/Visual Branch. Before returning to home waters the ships visited Cadiz (30 Apr.) where Queens Day was celebrated.

**** Federal Republic of Yugoslavia

TG passing under the Queen Juliana Bridge, Willemstad, Curaçao. Ceremonial entry in St. Anna Bay. Guide of the column turning to starboard, entering 'Schottegat'.

***Right:** Line of bearing: nearest to lens* Tromp, Jacob van Heemskerck *(F 812),* Jan van Brakel *(F 825), unknown S-frigate and* Amsterdam *(A 836).*

with *Van Nes* and *Willem van der Zaan*. They visited Kristiansand and then *Tromp* sailed to St. Petersburg to commemorate the 50th 'Victory Day' of the Soviet Union over Nazi Germany (5 - 10 May). After returning (13 May) the ship received more repairs.

In the first sailing week a fire broke out in the diesel-generator room (3 Sep.) It was extinguished rapidly, but the damage was done. It would take till 16 October before she would sail for work-up. 'Exercise-Ready'; from 6 - 8 November a fine visit to London and then to Plymouth where a group of 'Seariders' waited who were eager to give the Dutch frigate a 'shakedown'. The crew accepted the challenge of a tough examination for all departments. In the NOST***** (10 Nov. - 13 Dec.) *Tromp* scored 'Good' results. With a proud crew a warship 'Operational-Ready' entered Den Helder on 14 December.

(NL) ESKADER

(EKD - CEKD) / (NLTG - COMNLTG)

In NATO Task Organization: TG429.5 - CTG429.5 Throughout translated as Task Group (TG). Since 1 January 1996 indication changed in: 'Belgisch-Nederlands Eskader'. Even if no Belgian unit was present. To facilitate reading the indication will not be adapted.

The NLTG with Tromp. Witte de With (F 813), unknown S-Frigate and Zuiderkruis (A 832).

Left: Berthed in homeport with 8 Harpoon canisters. Most of the time equipped with only 4. If combat ready could carry 16 SSM's.

TG Cruise 1/96 (19 Feb.- 22 Mar.)
Flagship *Tromp* with Belgian Alouette embarked, *Jacob van Heemskerck*, *Bloys van Treslong* (+ Lynx), *Willem van der Zaan*, *Amsterdam*, *Westdiep* (19-25 Feb.) and *Van Galen* (18 - 22 Mar.).

Harbours: Bergen (23 - 26 Feb.) and Göteborg (14 - 18 Mar.) Ships were much hampered by very bad weather. Exercises on the program of 'Battle Griffin' had to be adapted frequently. On their way to Bergen the ships encountered such rough weather conditions, that most reported damage. While exercising even wind force 10 to 12 was reported and a Russian auxiliary ship popped up... Reason

enough to cancel the postponed missile launchings. The amphibious operation in Andfjord by the UK/NL landing force was successful.

TG Cruise 2/96 (21 Apr. - 31 May).
Tromp (flag, Lynx embarked), *Jacob van Heemskerck*, *Jan van Brakel*, *Pieter Florisz* (22 Apr. - 17 May), *Amsterdam* with Sea King-RN Squad. 810), *Witte de With* (11 - 31 May), submarine *Bruinvis* (S 810) and

***** Netherlands Operational Sea Training

Mercuur (A 900), also 2 NL Orion MPA's. On 15 May the TG anchored in Tagus river off Lisbon for a pre-exercise conference for Exercise Swordfish. On 29 May 3 FF's were invited to join FOST****** Weekly War. Harbours: Cadiz (29 - 29 Apr.), Lisbon (9 - 13 May), Leixoes (24 - 27 May).

TG Cruise 3/96 (24 Aug. - 21 Sep.)
Tromp with Lynx embarked, *Zuiderkruis* with 2 Sea Kings (Sqn 819) damaged by explosion in hangar; u/s (unserviceable). *Tjerk Hiddes, Abraham van der Hulst, Wielingen. Bruinvis, Mercuur, Witte de With* for a couple of days. Also P-3C Orions. Exercises:Northern Light/Bright Horizon (2 - 12 Sep.) and Falcon Nut (16 - 20 Sep.)

******   Flag Officer Sea Training

RN ships, 1 German and 2 Danish submarines, many planes and also 9 FPB's; in Falcon Nut a r/v with Spanish Baleares-class *Cataluña* (F 73).
Harbours: Copenhagen (30 Aug. - 2 Sep.) and Stavanger (12 - 15 Sep.).

When on 3 October the ship reached the age of 21 years, and became an adult, all former crewmembers were invited for the birthday party. For the occasion the characteristic dome was decorated with a beautiful bow tie in the national tricolour.

TG Cruise 4/96 (14 Oct.- 13 Nov.)
Tromp, Abraham van der Hulst (both with Lynx) *Jan van Brakel, Karel Doorman* (14 - 18 Oct), *Wielingen, Westdiep.*
Just before departure a short circuit caused

Left: Coming of age; dome decoration for 21st birthday.

Below: Tromp *and De Ruyter (R). Mainmast F 806 equipped with US Whiskey-3 (WSC-3) UHF SATCOM antennas of OE-82 Cartwheel type.*

a fire in the fore engine room dealt with by the watchkeepers. After r/v with *Wielingen* the TG proceeded to the Skagerrak. On 15 October a r/v with a German TG consisting of FGS *Bayern* (F 217),'brand-new' (comm. in May), *Emden* (F 210), *Augsburg* (F 213), *Glücksburg* (A 1414). The German CTG conducted exercise Autumn (14 - 25 Oct.). In the work up phase for JMC 3 the MCM units were hampered by heavy weather, an interesting surface action against SNFL******* was executed. The debrief of the (by weather disturbed) JMC 3 (28 Oct - 7 Nov.) was on 11 November.
Harbours visited: Faslane (25 - 28 Oct.) and Leith (7 - 11 Nov.).

TG Cruise 1/97 (10 - 21 Feb.).
Tromp, Witte de With, Pieter Florisz, De Ruyter, Bruinvis, Mercuur, Westdiep.
Springex 97 held in areas Skagerrak and Scottish Waters. Six navies executed ASW and AAW exercises. Torpedoes launched: *Bruinvis* (2), Orions (2), FF's (4) and one by a helicopter.

TG Cruise 2/97 (17 Mar.-26 Jun.).
Tromp, Zuiderkruis, Van Speijk, Willem van der Zaan, Westdiep.
"Westlant 97"was a demanding deployment. Training:

*******   SNFL = Standing Naval Force Atlantic

- OPS room officers class
- junior/senior midshipmen (in May)
- visit by students of Naval War College
- SAM and SSM missile and torpedo launchings
- exercises with "stationsschip" *Tjerk Hiddes*
- exercises with USS *The Sullivans* (DDG 68)
- exercises with DESRON 22 (USN) = USS *Laboon* (DDG 58), USS *Caron* (DD 970), USS *Simpson* (FFG 56) and also with USS *Providence* (SSN 719)
- exercises with German TG
- exercises with *Dolfijn*

On 22 April the TG passed the Tropic of Cancer. In May the CTG payed a visit to the National Pantheon in Caracas and placed a wreath at the resting place of Simón Bolivar (El Libertador).

Showing the flag by the ships and their - ±1000 'ambassadors'. Ports of call of *Tromp*: Halifax (Mar.), Newport, Baltimore, San Juan (Apr.), Roosevelt Roads, La Guaira, Willemstad, Fort de France (May), Dakar, Casablanca (June). The two last mentioned ports formed part of 'Maritime Presentation' and important authorities received a ceremonial welcome. The arrival in Den Helder concluded this successful multi

In the Norwegian fjords.

Ships of STANAVFORLANT seen from HMCS Toronto *(FFH 333). Left* USS Robert G. Bradley *(FFG-49), German Brandenburg class and* Tromp.

***Below:** BNLTG (Belgian - Neth. Task Group) conducting RAS with* Zuiderkruis. Tromp *and* Westdiep *by the 'standard' alongside method and* Willem van der Zaan *by astern method.*

mission cruise and was marked by a 'Fly By' executed by three helicopters...

10 June 1999, passing Hoek van Holland. Note tables for guests on the helideck 'terrace'.

On 13 September the Dutch COMSTAN-AVFORLANT embarked *Tromp* in Ponta Delgada. In Toulon (20 - 23 Sep.) preparing for Exercise Dynamic Mix in the Ionian Sea. Visit to Istanbul (9 - 14 Oct.). Exercises with STANAVFORMED 'en-route' to El Ferrol, *Tromp* had to return to 'De Schelde' for prop shaft repairs. Returned to El Ferrol (7 - 10 Nov.) then sailed with SNFL, visited La Coruña (21 - 24 Nov.) and Brest (28 Nov. - 1 Dec.). Staff transferred to new flagship HNlMS *Witte de With*. *Tromp* again sailed to her builders to repair the prop shaft troubles. On 13 December *Tromp* arrived in Den Helder taking over flagship duties the same day.

Den Helder, moored to jetty 18 in front of the Naval Fire Brigade barracks.

On 13 January 1998 *Tromp* departed Den Helder as flagship of SNFL and arrived in Lisbon (19 - 26 Jan.) to celebrate the anniversary of the formation of SNFL 30 years ago.
Working up with Portuguese units. Together with HMS *Argus* (A 135) *Tromp* carried out a SAR action. The Spanish

containership *Delphin de Mediterraneo* had been broken in two; 13 of the crew were rescued and transferred to Portuguese authorities. A visit to Malaga (5 - 9 Feb.) with 'Open House'. Exercising and port visits to Bordeaux (12 - 16 Feb.), Antwerp (20 - 24 Feb.) and Rotterdam. In Bodø (6 - 9 Mar.) the ships moored during a blizzard. A few days later the exercise Strong Resolve

commenced. *Tromp* sailed most of the time in Norwegian fjords. Endex at Trondheim (21 - 24 Mar.), 27 March entering Den Helder. Before arrival deck landing No 1000 was executed followed by a small celebration. On 3 April the command of SNFL shifted to the German Rear admiral embarked in FGS *Bayern* (F 217), *Tromp* was relieved by *Jacob van Heemskerck*.

In background 'Dok VI' (length inside 503 ft (153 m). On 9 June 1980 docking trials by Groningen *(D 813) and 23 June the first docking by* Tromp. *Officially put into use on 2 October by the Defence Minister.*

Along with combat supply ship *Zuider-kruis*, *Tromp* departed on 2 June for exercise Baltops. Visiting Gdynia, Warnemünde, Rostock, Kiel and Helsinki. Once returned the ship participated in the Navy Days (10 - 12 Jul.).
In Aug. sailing with guests of PR Dept. Later qualifications for flight deck officers and also sparring partner for *Bloys van Treslong* (F 824) which was preparing for FOST. The crew list was reduced to 179 on 1 October 1998 and 13 October the Tartar launcher was hoisted and stored ashore. Subsequently the housing case, deck module and antenna of SMART-L were placed. Winter leave commenced on 11 December.

When *Tromp* sailed on 21 January 1999 she showed an outstanding silhouette caused by the SMART-L radar antenna. The first port of call was Newcastle (22 - 24 Jan.).

Tromp assisted *Witte de With* which was preparing for the NOST. Bergen was visited from 12 - 15 Feb. During an extended visit to London (15 - 26 Mar.) *Tromp* was berthed alongside the museum-ship, cruiser *Belfast* to proudly present the

new radar to British experts and RN Staff and specialists. Thereafter *Tromp* assisted *Jan van Brakel* with a "Staff Sea Check" important for the intended Sea Training by FOST. (Later in May *Jan van Brakel* earned in the NOST a "VSAT"= Very Satisfactory=.

1998, 2 December = Tromp = *SMART-L radar (7800 kg) fitted for testing and trials.*

SMART-L

SMART-L = name of the 3D VSR (Volume Search Radar)
Electronic stabilisation - simultaneous tracking capacity of 1000 air-targets and 100 surface targets. In Dec. 1997 four systems were ordered for the new AAW-and Command Frigates. SMART-L =Signaal Multibeam Acquisation Radar for Targeting in the L-band.

Norwegian visits Bergen (3 - 4 Apr.) and Oslo (6 - 9 May). Tests/trials program finished; the SMART-L was dismounted on 20 May. Tartar was not installed again because the ship was to be decommissioned in November.

After some PR days in June the ship departed with CFREGRON embarked for a midshipmen cruise from 14 June to 16 July in company with *Witte de With*, *Van Galen* and *Wielingen*. Visits to the ports of Lisbon, Las Palmas, Cadiz and Leixoes. The end of the cruise was marked by a sailing Family Day. From 30 August Tromp sailed some days in behalf of the Maritime Heli-

1 November 1999. HNLMS Tromp, *arrival homeport to be decommissioned on 12 November. Flaghoist:"I am leaving formation-my duty is completed" Note: Mk. 13 launcher removed. After refurbishing by RR the engines will be sold to Brazil. Will be BU for scrap (spare parts for sister* De Ruyter). *Commissioned in 1975 she sailed over 500.000 nm. (Log recording: 510.800 nm.)*

Former commanding officers welcome Tromp *on her last arrival.*

copter Group. The flight deck utilized by pilots to qualify for their certificate. *Tromp* berthed in Rotterdam (3 - 6 Sep.) to brighten up the World Harbour Days.

In September and October carried out some PR trips and again supported the Lynx pilot qualifications. Departed 18 October bound for Dublin (23 - 24 Oct.). HNLMS *Tromp* paid farewell to the UK- our faithful & reliable ally- by a visit to Southampton (30 - 31 Oct.).
On 1 November the good ship entered Den Helder with a shark's mouth on her bow. To comply with ceremonial protocol a salute of 15 rounds was exchanged with the saluting battery ashore.
ADMIRALNLFLEET Vice Admiral

L.L.Buffart was welcoming and also former CO's, amongst them the present CINCRNLN Vice Admiral C.van Duyvendijk.

Decommissioned 12 November 1999. All hands mustered on the heli-deck to be adressed by the Captain. The ceremony was witnessed by former CO's and crew members. When finished the CO ordered to lower the Ensign, Jack and Commissioning (War) Pennant.
On 17 Dec. 2002 the dismantled ship was towed away from jetty 3 by the tugs *Barracuda* (2700 hp) and *Rotte* (1600 hp) to the shipbreaking company in 's Gravendeel. The breaking up started in September 2003.

Series editor	Contributors
Jantinus Mulder	Luuk Kuilder
	Capt, Bob Roetering
Publisher	(RNLN rtd)
Lanasta	Dick Vries
Authors	**Graphic design**
Jantinus Mulder	Jantinus Mulder
Henk Visser (FCCY rtd)	

First print, October 2021
ISBN 978-90-8616-402-8
NUR 465

Mail address: Warship:
Oude Kampenweg 29, 7873 AG Odoorn
The Netherlands
Tel. 0031 (0)591 618 747
info@lanasta.eu

Lanasta

Awaiting her final destination.

Former CO's were offered a letter of the name-board

Right:
Bridge/radardome of De Ruyter *and Bofors 12 cm guns of* Tromp *placed on the grounds of the NL Naval Museum.*

References

- *Conway's All the worlds Fighting Ships 1922-1946, Roger Chesneau (editor), 1980 Conway Maritime Press*
- *Ir. Frans O.J.Bremer*, Radar Development in the Netherlands, *Thales Nederland 2004*
- *Norman Friedman*, The postwar naval revolution, *1986, Conway Maritime Press*
- *John Jordan*, Tromp, Warship Vol III -No.9, *Conway Maritime Press 1979*
- *Mark Chris*, Geleidewapen Fregatten Hr.Ms. Tromp en Hr.Ms. De Ruyter
- *Capt. S.G. Nooteboom*, Deugdelijke schepen, Marinescheepsbouw 1945-1995; *Europese Bibliotheek (Zaltbommel, 2001)*
- *Capt. J.H. van der Veen*, Machinery installation.
- Jaarboek Koninklijke Marine (*various volumes*)
- Warship World (*various issues*)
- *Website: Dutchfleet.nl*